ADVANCED [illegible]L

Practical Work for BIOLOGY

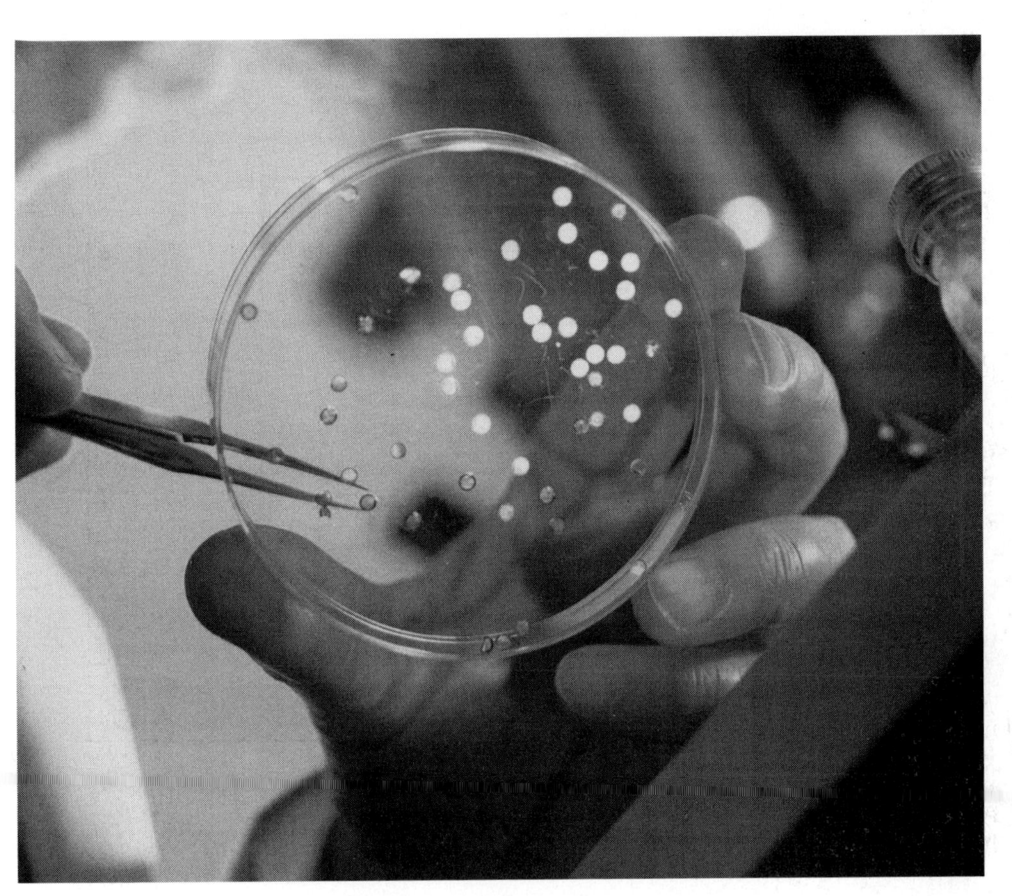

Sally Morgan

Acknowledgements

The publishers would like to thank the following individuals, institutions and companies for permission to reproduce photographs in the book. Every effort has been made to trace ownership of copyright. The publishers would be happy to make arrangements with any copyright holder whom it has not been possible to contact:

Biophoto Associates page 60; Andrew Lambert, pages 6 and 9.

Orders: please contact Bookpoint Ltd, 130 Milton Park, Abingdon, Oxon OX14 4SB. Telephone: (44) 01235 827720. Fax: (44) 01235 400454. Lines are open from 9.00–6.00, Monday to Saturday, with a 24 hour message answering service. Email address: orders@bookpoint.co.uk

British Library Cataloguing in Publication Data
A catalogue record for this title is available from the British Library

ISBN 0 340 84712 3

First Published 2002
Impression number 10 9 8 7 6 5 4 3 2 1
Year 2007 2006 2005 2004 2003 2002

Typeset by J&L Composition Ltd, Filey, North Yorkshire.
Printed in Great Britain for Hodder & Stoughton Educational, a division of Hodder Headline Plc, 338 Euston Road, London NW1 3BH by Martins The Printers, Berwick upon Tweed.

CONTENTS

INTRODUCTION

Every biology course includes some practical work. It is an essential part of understanding the principles of biology. There are a number of 'classic' experiments that all biology students are expected to have seen and which may be examined in the written examinations. In addition to the class practical work, biology students studying AS and A2 biology are required to carry out coursework or formal practical tests that will be assessed. The marks from these assessments go towards the overall grades.

The new biology specifications set out the subject criteria. There are four main assessment objectives (AO). They are: AO1 Knowledge with understanding; AO2 Application of knowledge and understanding, analysis, synthesis and evaluation; AO3 Experiment and Investigation; and AO4 Synthesis of knowledge, understanding and skills.

Within AO3 Experiment and Investigation, students are expected to be able to:

- devise and plan experiments
- demonstrate safe and skilful practical techniques
- make observations and measurements and record them accurately
- interpret, explain, evaluate and communicate the results of their experiments using their biological knowledge.

The practical coursework is an important element of the A level examinations. At AS and A2 level it represents between 15 and 20% of the final marks.

The awarding bodies in England and Wales assess students on four main skills: planning, implementation, analysis and drawing of conclusions and evaluation. A detailed examination of these skills forms the bulk of the information presented in this book. There is one chapter devoted to each of the four skills. Within each chapter, the skill is examined and information provided on how to approach the assessment, together with helpful hints. Each of these chapters ends with an example of a piece of student's work. This is accompanied by a commentary which indicates the strengths and weaknesses of the work and how it could be improved.

Although the coursework requirements of the four awarding bodies is generally very similar, there are a few differences. Therefore, Chapters 8 to 11 examine the specific criteria set by each awarding body.

1.1 Progression from AS to A2

The skills that are examined at A2 are very similar to those examined at AS. However, students will find that their practical skills at A2 will be tested at a higher level than those tested for AS. For example, students will be expected to discuss and analyse their findings in greater depth and to be able to carry out a statistical test. The A2 coursework often requires students to carry out an individual study. Chapter 7 focuses on individual investigations. It examines some of the skills involved in the assessment and provides ideas for possible investigations.

1.2 Mathematical skills

Whenever data is collected and analysed, mathematical skills are usually involved. Students are expected to be able to use ratios, fractions, percentages and carry out basic mathematical calculations to find the mean values and determine a rate of change. Students should be able to present their data in tables and graphs and where necessary carry out simple statistical tests. Within this book, students will find guidance on the manipulation and presentation of data and use of statistical tests.

The mathematical skills that are expected are summarised below. You will be expected to be able to:

- recognise and use expressions in decimal standard form
- use ratios, fractions and percentages
- make estimates for the results of calculations without using a calculator
- use a calculator to find the sum, inverse ($1/x$) and the square root
- find an arithmetical mean
- use an appropriate number of significant figures
- translate information between graphical, numerical and algebraic forms
- construct and interpret frequency tables and diagrams, bar charts and histograms
- plot a graph showing two variables using data obtained from an experiment or other source
- calculate the rate of change from a graph showing a linear relationship
- **understand probability to work out genetic ratios**
- **understand the principles of sampling**
- **understand the importance of chance when interpreting data**
- **use a scatter diagram to identify a correlation between two variables**
- **use a simple statistic test.**

(the bold points are for A2)

At the end of the book you will find a glossary covering the terms relating to biological coursework and an appendix with useful addresses.

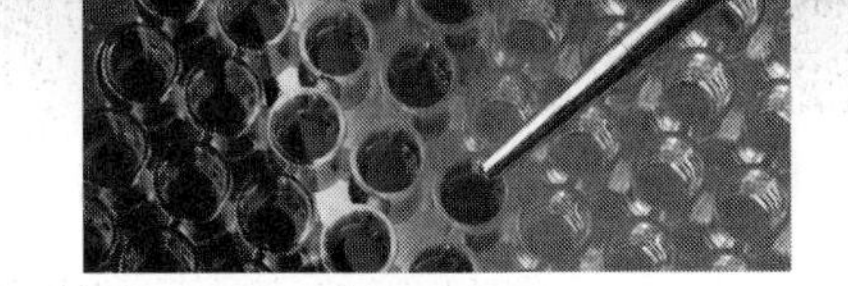

CHAPTER ONE

Planning

1.1 Key elements of planning

You should be able to:

- identify and define the nature of a question or problem using available information and knowledge of biology
- choose effective and safe procedures, selecting appropriate apparatus and materials and deciding the measurements and observations likely to generate useful and reliable results
- consider ethical implications in the choice and treatment of organisms and the environmental and safety aspects of the proposed procedures.

Your ability to plan an investigation will be tested at both AS and A2 levels. You will be expected to use the information you have learnt in lessons, together with information obtained from secondary sources such as textbooks, library books and on the Internet. The plan that you prepare should describe in detail all the stages of the investigation, including the apparatus and procedures that you intend to use and the types of measurements that you will make. You will also be expected to make a risk assessment.

1.2 Research

The freedom to choose any subject of your choice for investigation will probably depend on whether you are carrying out an AS or an A2 investigation and the availability of resources and apparatus at your school. You may find that your teacher selects a topic to be investigated and students design their own investigation based on that topic, for example you may be restricted to investigating enzymes or to a particular unit or module of the specification. Sometimes the time of year will restrict the range of investigations you could carry out.

Having decided on the topic for investigation, you need to start your research. The first place to start is your course notes and textbooks. Other sources include information books in the libraries and Web sites on the Internet. For example, you may be interested in a particular plant species that is found growing in your school grounds. Research can provide useful information about where you can expect this plant to grow, whether it prefers a particular soil type or pH, dry or wet ground, shade or sun. It may have an important relationship with another plant or an animal species, for example it may be dependent on a certain insect to pollinate its flowers. It could have an unusual method of seed dispersal or certain animals could feed on it. If you are planning an enzyme investigation you would need to research the conditions under which the enzyme works well, its substrate and the products of the reaction. You may want to discover whether the enzyme is used commercially. All this information is useful when you are planning an investigation.

1.3 A testable hypothesis

The next stage is to produce one clear, well-defined, testable hypothesis. A hypothesis is a bit like a prediction or 'educated guess'. Using the information generated by your research, you start by making a predication about the results. Then you carry out the investigation and use your actual results to prove or disprove your hypothesis. Consider the following hypothesis: A beech leaf will lose more water through transpiration than a laurel leaf. An investigation is designed to test this hypothesis. If the results support this hypothesis, the hypothesis is accepted. If on the other hand, the results show that there is no difference in the water loss, you can reject your hypothesis.

Many statistical tests require you to set a null hypothesis. A null hypothesis would state that there is no difference in the water loss from a beech leaf and a laurel leaf. If you discovered that there was a difference in water loss, the null hypothesis would be rejected. (See Chapter 5 on statistics.)

As the term suggests, a testable hypothesis is one that can actually be investigated.

Some students find it difficult to produce a testable hypothesis. Don't worry if you can't do this on your own, as you can ask for help, but this will be reflected in your final marks. The important thing is that you produce a hypothesis that can be tested. Without this, the rest of the investigation will be meaningless and you may lose marks.

The hypothesis should be accompanied by a brief paragraph that outlines the background biological information to support the hypothesis. Make sure that the background information that you provide is relevant to your hypothesis. Often, it is very tempting to add all sorts of information that looks good but does not add any value to your plan.

1.4 Variables

Once you have a hypothesis, you can move on to the next stage in the design of the investigation. A critical part of your design is the consideration of the variables. There are two types of variable: the independent variable and the dependent variable. Your plan needs to show how you are going to control the independent variables. The independent variable is the factor that is being controlled, whereas the dependent variable depends on the independent variable. Start by making a list of all the variables that could affect your investigation. Which one are you going to change? For example, in an enzyme investigation you could study the effect of changing the amylase concentration on the conversion of starch into maltose. The amylase concentration is the independent variable that you are going to change. Your dependent variable is the concentration of maltose. You will monitor how the concentration of maltose changes during the investigation. At each concentration of amylase, you will measure the quantity of maltose that is produced. All the other variables that could affect the enzyme reaction have to be controlled.

One other point to consider is the range of values for your independent variable. For example, if you are going to vary the temperature at which an enzyme reaction takes place, over what range of temperatures are you going to make measurements? Do you want to study the effect over a wide range or focus on a narrow range? Ideally, you would want to carry out the investigation at a number of different temperatures, for example 10, 20, 30, 40, 50°C. If you think the optimum point will be around 40°C, there's not much point carrying out the investigation at 10°C. It would be better to have more points in the middle of the range, for example 30, 35, 40, 45, 50°C. Remember that you will probably produce a graph, so you will need a reasonable number of points to plot on the graph. The investigation will have to be repeated at each temperature and this will affect the length of time that you will need to spend on the investigation. All these points have to be taken into consideration at the planning stage.

All the other variables, such as temperature, pH, concentration and volume of substrate, have to be controlled in order to produce reliable results. They must not vary during the investigation otherwise the results will be affected. Your plan must include some information on how you intend to control these variables. Some will be relatively straightforward, for example use the same volume of substrate throughout the investigation. However, others will require more careful management. You cannot assume that the variables will stay constant, so you will need to show how you will monitor them during the investigation. For example, the temperature of the water bath may rise slowly if it is placed close to a light source, so you would need to take regular temperature readings. If the temperature changed, you would have to take action to bring the temperature back to the correct level.

Many investigations require the use of a control. In the enzyme investigation described above, a suitable control could be one treatment in which the enzyme is replaced by distilled water. Not all investigations can use a control, but it is important to include one where appropriate. Every plan should have a comment about the use of a control, even if it is just to state that it was not possible to have a control.

1.5 Statistics

The design of your investigation will affect the type of statistical test that can be used on the results. If you are going to include a statistical test, you need to decide at an early stage which test you will carry out and design the investigation appropriately. There is more information about designing these types of investigation in Chapter 5.

1.6 Reliability, accuracy and precision

Once you have a testable hypothesis and you have identified the variables, you need to think about the apparatus and the procedures that you are going to use. The apparatus and procedures should be suitable for you to take reliable and accurate measurements.

Reliability is defined as the degree or measure of confidence that can be placed in a set of observations or measurements. Put more simply, are the results trustworthy? The reliability of a set of results depends on the number of observations or measurements that were taken and the accuracy of each observation or measurement. Accuracy is defined as the degree to which a measurement approaches the true value. Another term which is used in connection with accuracy is precision. Precision is that part of accuracy that lies within your control. It involves your choice of apparatus and the skill with which you use it.

The first step towards accurate measurements is your choice of apparatus. For example, you are going to monitor the changes in pH of a solution. There are a number of options available to you. You could use a piece of universal indicator paper. You dip the paper into the solution at regular intervals and compare the colour changes with a series of colours on a chart. This method is a bit 'hit and miss' as it is often difficult to match the colours, so you may have introduced a degree of error into your results. A far more reliable and accurate method would be to record the changes using a pH meter or a data logger. Similarly, you may need to measure masses during your investigation. There are several types of balance including the top-pan balance, the spring balance and the lever balance. You need to choose a balance that will allow you to weigh out masses to the appropriate level of precision.

There are a number of pieces of equipment that you could use to make measurements of length. They include a tape measure, a metre rule, a 30 cm ruler, a vernier caliper (Figure 1.1) or a graticule in the eyepiece of a microscope. The choice of equipment depends on what you are measuring and the precision with which you want to make the measurements. A tape measure will be fine for determining sampling positions along a belt transect in an ecology investigation, but you wouldn't want to use it to measure the lengths of leaves. You would want to use a more accurate instrument to measure lengths and diameters of small objects such as a seed. A metre rule is used to measure to the nearest millimetre, while a vernier caliper measures to the nearest tenth of a millimetre. In general, the number of significant figures gives an indication of the precision of the measurement. For example a measurement on a metre rule is 5.0 ± 0.1 cm, while a vernier caliper reading would be 5.0 ± 0.01 cm. Another point to consider is the accuracy of the ruler itself. It is likely that an expensive steel metre rule has more

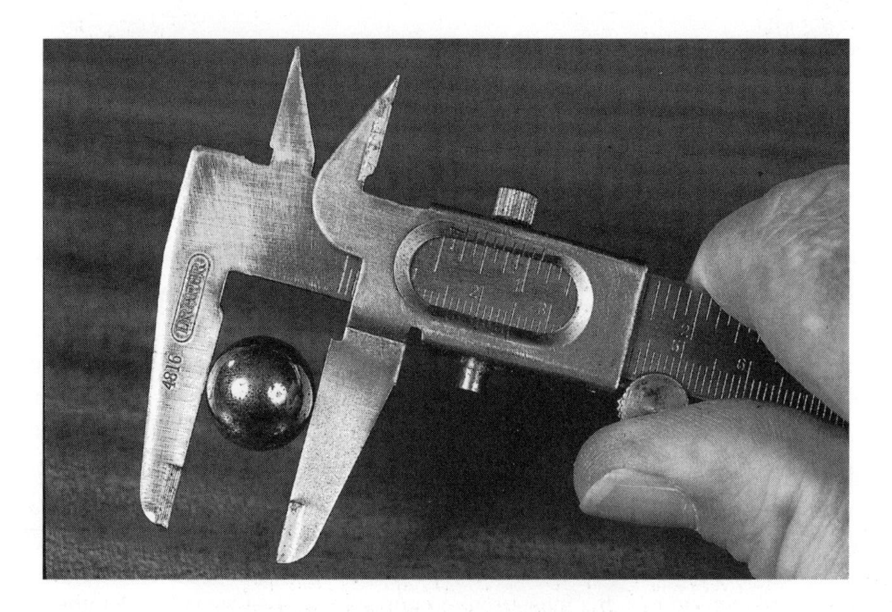

Figure 1.1 A vernier caliper for making accurate measurements

accurate markings than a cheap wooden or plastic metre rule. The same source of error applies to the shorter rulers. To improve the accuracy and precision of your measurements, you may decide to use a vernier caliper to measure objects up to 100 mm in length. However, skill with which you use the apparatus will also affect the accuracy of the measurements. There is little point using a vernier caliper if you cannot use it properly, as you would introduce too much error into the measurements. You would be better to use a good quality steel ruler.

A number of investigations involve the measurement of a volume, for example the volume of oxygen evolved by a catalase reaction. The simplest but least accurate method would be to carry out the reaction in a cylinder and measure the height of the froth. Alternatively, the oxygen gas could be trapped in an upturned test tube and the volume estimated. These methods would not gain you much credit as they are typical of the methods used at Keystage 4 and you need to be able to show some progression. Instead, you would need to consider a method that allowed you to measure the volume of oxygen accurately. For example the oxygen could be collected in a graduated tube. Another method would be to use a data logger to record the changes in pressure that occurred during the reaction (see Figure 1.2).

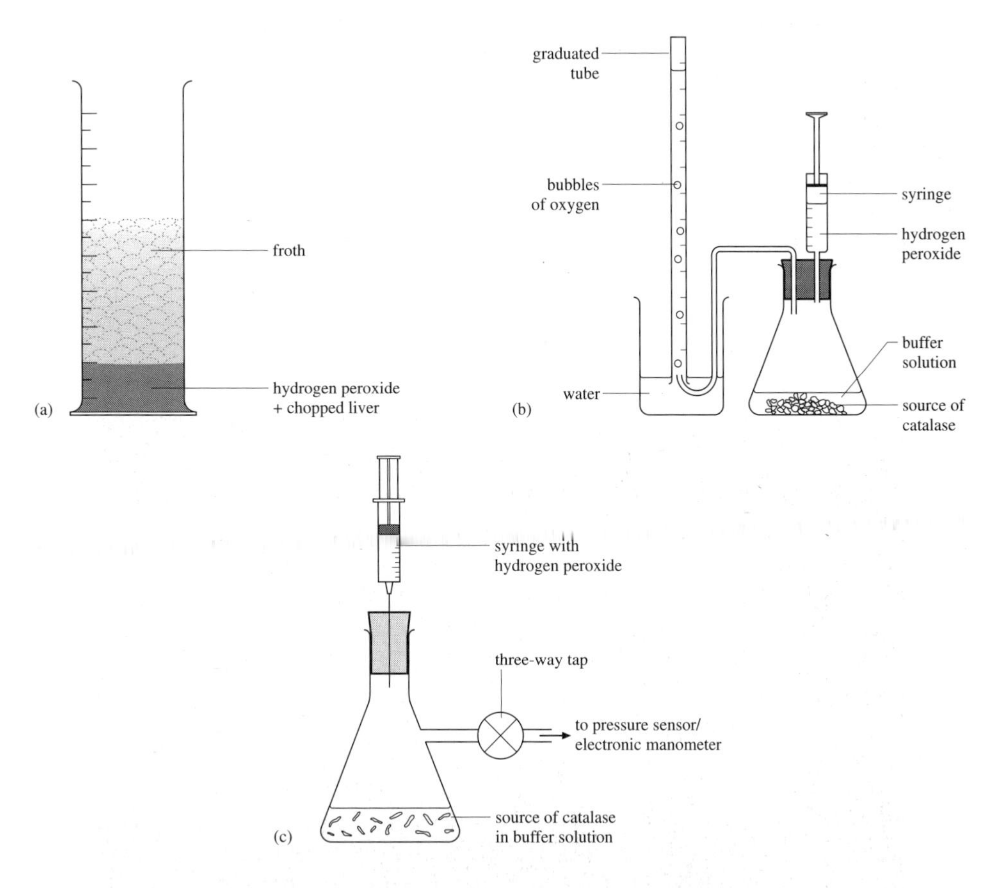

Figure 1.2 *There are several ways of measuring the volume of oxygen gas released by the decomposition of hydrogen peroxide in the presence of catalase:* ***(a)*** *measuring the height of froth using a cylinder,* ***(b)*** *measuring volume of oxygen gas in a graduated tube,* ***(c)*** *recording the pressure changes that occur when oxygen gas is evolved using an electronic pressure sensor*

Part of the implementing assessment (see Chapter 2) requires you to demonstrate a wide range of manipulative skills. You should bear this in mind when you decide on the techniques that you will use in your investigation. Remember to indicate why you have chosen certain pieces of equipment.

Once you have decided on the most suitable apparatus, you need to make a list of all the pieces of apparatus that you will require during the investigation, including the numbers of test tubes, labels etc.

1.7 The frequency of measurements

Now you have planned what you will measure and how you will measure it, you have to decide on the frequency of your measurements.

The frequency of your measurements can be determined by how long it takes you to actually make the measurement. For example, you may have to take a draw off a sample before you make the measurement and this could take many seconds. You may not have time to make a measurement every minute. In this case it would be better to specify that you make one measurement every 5 minutes.

You have to decide the length of period over which you will take measurements. Will you take measurements for 20 minutes or for 60 minutes? One set of data does not tell you much, so you will need to repeat the measurements a number of times in order to obtain mean (average) values (see below). The length of time over which you make measurements will limit the number of replicates that you can carry out. If each set of data takes 1 hour to obtain, then it will only be practical to run the measurements three or four times. If the measurements run for 20 minutes, then you may be able carry out more replicates (repeat measurements).

1.8 Reliability and the number of replicates

One set of measurements is not usually very reliable. For example, the height of one person in your class would not be very representative and would give no indication of the range of heights. Measuring the height of more people in the class and calculating the mean value would give you a more reliable value. In order to obtain reliable results it is necessary to carry out the investigation several times. As indicated above, the type of investigation and the length of time it takes to obtain measurements will determine the number of replicates that you take. Obviously, the more replicates, the more reliable your results. Often it is difficult to obtain a large number of replicates when carrying out ecological investigations, so two or three replicates are quite normal. Laboratory investigations, especially those involving enzymes, can be carried out more quickly and so it is more likely that you could get five to ten replicates. However, there is a limit to the number of replicates you should make. There is no point in making 20 or more replicates. As well as wasting time, the extra sets of data probably won't add any extra value to the overall results. More information concerning the choice of number of replicates is given in the chapter on statistics.

1.9 Risk assessment

By now you should have a draft plan. At this stage it is important to consider the risks of your proposed procedures. Some of the common risks include:

- using a Bunsen burner to heat something
- handling chemicals that are described as being corrosive, irritant, toxic or harmful
- using sharp instruments such as scalpels.

For example, if you are carrying out an enzyme investigation, you may want to use a water bath to control the temperature. If you use a beaker and tripod over a Bunsen burner, there is a chance that you could knock the beaker and spill warm water. It would be safer to use a thermostatically controlled water bath. Check the labels on the chemicals you intend to use. The label will display symbols indicating how the chemical should be handled (Figure 1.3). For example, if you are going to use a corrosive liquid such as an acid or hydrogen peroxide you should use safety goggles and wear a lab coat.

Figure 1.3 *The label on this bottle of dilute nitric acid warns that the chemical is corrosive*

When preparing your risk assessment, think about the main risks in using a substance or carrying out a particular procedure. See if you could use something else to avoid the risk completely. If this is not possible, then consider what you could do to reduce the risk to an acceptable level. You may find it useful to prepare a risk assessment form in which you identify the nature of the hazard, the source of your information, how you will control the risk and any emergency procedures that are necessary. For example, what should you do if you spill a dangerous chemical on your skin or on the bench?

1.10 Ethics and the law

If you are designing an investigation that requires the use of living organisms, it is important that you consider whether it is necessary to use living organisms in the first place. If you do have to use them, then you must ensure that the treatment they receive is acceptable. If you are carrying out an ecological investigation, then you must consider the effect your investigation could have on the environment. Does it disturb animals? Will you have to trample vegetation in order to take your measurements? Are there any long-term effects of taking samples from an area?

Plants

Whole plant specimens should not be removed from their natural environment for whatever reason. Most wild plants are protected by various Acts of Parliament and should not be picked or uprooted. If your proposed biological work requires the use of leafy stems and flowers, you should ensure that they are obtained from gardens or from commercial sources or bred for the purpose. Care must be taken during ecological investigations that plants are not trampled or damaged during the sampling process.

Invertebrate animals

No experiments should be carried out which may cause harm or suffering to animals. In no circumstances should the animals be subjected to such extreme conditions that they are killed. For example, experiments which investigate the effect of alcohol on *Daphnia* should use concentrations that do not kill the organism. It is not acceptable to carry out experiments that measure the amount of force needed to dislodge limpets or other invertebrates from rocks.

Any animal removed from its environment for the purpose of counting or measurement should be returned as soon as possible. Particular care should be taken when sampling invertebrates found in rivers, ponds and on rocky foreshores. Some investigations may involve the use of 'mark/recapture' sampling. Before this is carried out, consideration should be given to the effects upon individuals and populations of the organism.

Some ecological investigations may involve the use of pitfall traps. These traps should be checked twice daily to ensure animals are not trapped for long periods of time. The animals should be given protection from the weather and also a source of food.

Vertebrate animals

Experiments should not be carried out using live vertebrates. The only possible exception to this is maze learning in rodents. These behaviour experiments should be carried out without any harm or suffering to the animals and in no circumstances should they be subjected to such extreme conditions that they are killed. They should not be exposed to any unnecessary stresses such as deprivation of food, water or rest. For example, in experiments using a food reward during maze learning in rodents, the animals must not be starved prior to the tests.

Care must also be taken during sampling of habitats, such as ponds, so that protected vertebrates are not handled or disturbed. For example, the great crested newt is protected by law and it is an offence to handle these animals.

Humans

Many biological experiments are carried out where the subject is one of the students in the school. It is not acceptable to carry out investigations where students are given cigarettes, alcohol or other drugs. All subjects should give their consent to any activities before they are undertaken. A person aged 16 or 17 may give consent, without parental permission, to an investigation that will not harm them, either physically or mentally.

Any investigation involving young people under the age of 16 must be carefully planned and must not involve any chance of harm. For example, when carrying out investigations into fitness and heart rate, it would be essential to determine that the subject was fit and healthy and did not have any medical problems such as asthma, which could put them at risk. In addition, careful consideration should be given before any investigations are conducted where the subjects may be exposed to judgmental comments by their fellow students. This includes exercise experiments, reaction times, memory tests and testing for various genetic phenotypes. Some investigations compare different age groups. Again, great care must be taken to ensure that none of the subjects may be upset by the results of the investigation, for example, slow reaction times or reduced memory in elderly people. Any investigation that requires subjects to fill in a questionnaire must ensure that none of the questions are of a personal nature and that there is always an option not to answer any or all of the questions if the person so desires.

It is recommended that the Head of Department or the Headteacher be informed of all investigations involving students and that full risk assessments are carried out.

1.11 Pilot study

In many cases, it is appropriate to carry out a small pilot study that will identify any unforeseen problems. For example, you may not be sure that your chosen method will work or whether the concentrations that you have chosen will produce measurable results. So you should carry out a small test to check that your method will work in the chosen time period and give results. Once you have carried out the pilot, you can make final adjustments to your method before starting the full investigation. The results of the pilot study should be included in your plan.

1.12 Presenting your plan

The most important aspect of your plan is that it is clear and easy to follow. The person reading the plan needs to know exactly what you intend to do. So keep the writing concise and accurate. It should be in the first person, almost as a set of instructions. Sub-headings will help the reader focus on the different areas.

Checklist for planning an investigation

★ Have you prepared a clear, straightforward testable hypothesis?
★ Does your hypothesis state what you expect to happen?
★ Have you explained the background knowledge that supports your hypothesis?
★ Have you identified all the variables that could affect your investigation?
★ Have you used a suitable range of values for the changing variable?
★ Are your variables discreet or continuous?

Checklist for planning an investigation *continued*

- ★ How will you control the independent variables?
- ★ Have you discussed the techniques you will use and included your reasons?
- ★ Have you listed all the apparatus you will use?
- ★ Have you stated all the quantities and concentrations that you will use?
- ★ Have you stated what you will be measuring or observing and how you will do it?
- ★ Have you considered the accuracy of your measurements?
- ★ How many measurements will you make?
- ★ Over what period of time will you carry out your measurements?
- ★ How will you record your results?
- ★ How many replicates will you need?
- ★ What controls will you use?
- ★ Do you need to carry out a statistical test?
- ★ Have you carried out a pilot study?
- ★ Have you completed a risk assessment?
- ★ Does the investigation involve the use of living organisms?

Example: Planning an investigation (AS level)

The plan below was written by a student taking AS biology. It is followed by a commentary, which looks at the plan in terms of coursework criteria. There are suggestions on how it could be improved. A plan should tell you everything you need to know in order to carry out the investigation. Once you have read this plan think to yourself: could I carry out this investigation without needing further guidance? Are you clear about the apparatus and the method? What measurements were to be made? Were any precautions taken? How would you have approached this problem?

An investigation into the effect of wind on the rate of transpiration of privet leaves

Prediction

I predict that there will be more transpiration from leaves in the presence of wind. Secondly, the shoot with the smaller waxy leaves will lose less water in both still and windy conditions.

In the control test where there will be no wind, water vapour will be able to build up in the air spaces of the leaf and form a layer around the leaf as water transpires out. This will reduce the water potential gradient between the inside of the leaf and the outside. This will reduce the rate of transpiration. If wind blows across the leaves, this layer of water vapour will be blown away. The water potential gradient between the inside of the leaf and the air outside will increase and so the rate at which water transpires into the air will increase. Leaves can lose water through their upper epidermis but they generally lose more through their lower epidermis where there are more stomata. Plants that live in dry conditions tend to have thicker leaves with a smaller surface area and a thick waxy cuticle on their upper epidermis. These features reduce transpiration.

Background information

Transpiration is the loss of water vapour from the surfaces of a plant. Solar energy turns the water in the plants into a vapour causing it to evaporate into the leaf's internal air spaces before diffusing out of the stomata into the air.

As the water evaporates out of the top of the plant, it creates a suction on the column of water below it in the xylem. The upwards force on the column of water created by transpiration and the downwards force due to gravity creates a tension in the column of water. As the upwards pull is greater than the downwards pull, the column of water moves up the xylem. It is the evaporation of water from the leaves which causes the upwards movement of water. The water molecules have a high cohesion as they are polar. The water molecules are held together by weak hydrogen bonds.

The water is able to evaporate out of the leaf as the leaf has a high water potential and the air has a low water potential. The water molecules pass down the concentration gradient from the spongy and palisade mesophyll cells into the leaf's internal air spaces before diffusing out into the air.

Transpiration is needed to keep the cells of the spongy and palisade mesophyll cells moist, as this allows carbon dioxide to dissolve before diffusing into the cells for photosynthesis. The stomata open in the day to let carbon dioxide diffuse in and oxygen diffuse out, as part of photosynthesis. At night photosynthesis is unable to take place due to the absence of light so the stomata are closed to reduce water loss.

Light causes potassium ions to be pumped into the guard cells, which lowers their water potential and so water diffuses into the guard cells causing them to go turgid and so open. At night potassium moves out of the guard cells into the surrounding cells, so the water diffuses out of the guard cells causing them to close.

Apparatus

A fresh privet shoot (privet has waxy leaves) and a fresh beech shoot (the leaves are thin and do not have a waxy cuticle)
potometer
desk top fan
scalpel
bowl of water
stand and clamp
stop watch
ruler
vaseline

Method

1 Place the privet shoot under water in a large bowl and cut off the bottom 2 cm of the stem using a scalpel. I will do this to remove any blockages that are in the xylem from when the shoot was originally cut. Leave the cut end of the privet shoot under water to stop any air bubbles getting into the xylem.

2 Fill the potometer with water by submerging it in a bowl of water. Make sure there are no air bubbles in the capillary tubing.

3 While the potometer and shoot are still under water, insert the cut end of the shoot into the end of the capillary tubing. Make sure no air bubbles get in. Seal the join between the tube and shoot with vaseline.

4 Remove the potometer and shoot from the bowl and place in an upright position. Use the clamp and stand to hold everything in position.

5 Leave the apparatus to equilibrate for about 5 minutes.

6 The first experiment is the control. This test will give the rate of transpiration under normal inside conditions. These results form a control set against which the other results can be compared. Leave the apparatus in a bright position under a bench lamp and make sure there is no wind. Introduce an air bubble into the system. Measure how far the bubble travels along the capillary tube every 30 seconds for five minutes.

   When measuring the distances, I will make sure that my eye is at a 90° angle from the bottom of the bubble. This way I will avoid a parallax error when looking at the scale on the ruler. I will take all my measurements from the bottom of the bubble in order to ensure consistency in the measurement taking.

7 This experiment will be repeated five or six times so that an average of the rate of water uptake can be obtained. After each experiment, I will refill the tubing by releasing water from the reservoir and remove any air bubble in the capillary tubing.

8 The air temperature of the room should also be noted. This should be constant throughout the experiment.

The experiment will be repeated with different environmental conditions. I will see how wind affects transpiration by using a fan to blow cold air onto the leaves. The fan will be placed at a set distance from the leaves and left running on the same setting. I will make sure that the fan does not buffet the leaves as this could cause the stomata to close.

The wind experiment will be repeated several times in order to obtain more accurate data and to get rid of any anomalies that may occur in a single experiment. I will note the results for each 30-second period in all the experiments.

Then I will repeat the experiment, replacing the privet shoot with the beech shoot.

Once I have all the measurements I will work out the rate of water uptake for each 30-second period for all the experiments. Then I will plot a graph showing rate of water uptake against time without wind and with wind.

I can also work out the volume of water taken up per minute. I will use the following equation:

$$\text{volume of water uptake} = \pi \times \text{radius of the capillary tube}^2 \times \text{distance travelled}$$

Finally I will work out the surface area of the leaves on the shoots. I will do this by drawing around each leaf on a piece of graph paper. I will calculate the area by adding up the number of squares and then doubling the final number to give the total surface area in cm^2.

Risk assessment

There are no safety problems with this experiment. I will take care when using the scalpel, especially when using it underwater. I will make sure that water is not dripped on the fan, which is an electrical appliance.

Teacher's comments

There are a number of problems with this plan.

There is a single shoot of privet and beech and they are used through out the experiment. No guidance is provided on how long the shoot should be and whether the number of leaves on each shoot should be the same. Once the apparatus has been assembled, it is placed on the bench beside a lamp, but no information is given about the distance between the apparatus and the lamp nor the wattage of the lamp. It is unclear how the bubble will be introduced. The method of resetting the apparatus is also unclear. The apparatus is equilibrated for about 5 minutes. 'About' is not really accurate – is it exactly 5 minutes or 5½ minutes? It is unclear if there is any period of equilibration between the different experiments. Taking a reading every 30 seconds seems a lot. The student may not have time to make a reading and work out the distance travelled within 30 seconds and be ready to take the next reading. One-minute intervals would probably be easier. Again there is some inaccuracy concerning the number of times the experiment is repeated. It states five or six initially and later states 'several times'. The plan needs to be exact. The second part of the experiment involves using a fan to create wind, but there are no details about the type of fan, the speed at which it is set or how long it is left running before taking the readings.

There are no details on how the rate of water uptake will be calculated from the raw data. Also the calculations for the volume of water are based on water uptake per minute, while all the readings are water uptake per 30 seconds. Finally, the plan requires the student to work out the surface area of the leaves, but no information is provided on what will be done with this data once it has been calculated.

Overall, this is rather basic experiment which with a little more thought could be greatly improved.

Now consider this plan against the criteria set by the examination boards. The prediction (hypothesis) is testable, but it has been confused by including two predictions, rather than one clear prediction. Biological information is provided, but it does not really explain the nature of the problem. This experiment is looking at how two factors affect transpiration – the type of leaf and wind, but these factors are not really discussed in the biological background material. A potometer actually measures the rate of water uptake by the shoot, not the actual rate of transpiration from the leaves, although the two are linked. There should be some explanation of this in the introductory information.

The plan is based on a familiar procedure, although some new variables have been introduced. Two independent variables have been clearly identified – leaf type and wind, but there is little further detail. There is a brief mention of keeping all the other variables constant (light and temperature), but there is no detail on how this will be achieved. The student has described how the measurements for the dependent variable will be taken and these should generate reliable results. More information is needed on how the data on water uptake will be linked to the data on leaf area.

The apparatus and procedures are described, but more detail would have been helpful. For example, in describing how the experiment would be reset and setting up the fan. A risk assessment has been made.

Overall, this is a relatively low scoring plan. More marks could be obtained by adding more detail, for example detailing how the variables would be controlled, using a greater range of plants and varying the wind strength.

CHAPTER TWO

Implementation

2.1 Key elements of implementation

You should be able to:

- use apparatus and materials in an appropriate and safe way
- carry out work in a methodical and organised way with due regard for safety and with appropriate consideration for the well-being of living organisms and the environment
- make and record detailed observations in a suitable way, and make measurements to an appropriate degree of precision, using Information Communication Technology, where appropriate.

This chapter examines the criteria that are associated with implementing or carrying out investigations. You will be assessed on the way you carry out an investigation, your skill in making measurements and using the apparatus.

2.2 Working safely

You will be assessed on the way you use your chosen apparatus and carry out the investigation. There are some simple rules to follow to make sure that you work safely in the laboratory.

- Always move slowly around a laboratory.
- Wear a protective lab coat and goggles where necessary.
- Tie long hair back and do not wear dangly necklaces and earrings.
- Tuck your bags and other personal items safely under the bench before you begin.
- Never place your fingers in your mouth or eyes after using chemicals or touching biological specimens.
- Make sure that fragile objects such as glassware are placed where they cannot be knocked over or rolled off the bench.
- Allow hot objects such as Bunsen burners, tripod, gauze and beakers to cool down before you handle them.
- If you are not sure how to do something, ask for help.

You will have written a risk assessment as part of your plan. Make sure you read your risk assessment before starting the investigation and that you take adequate precautions to minimise any risks. Your risk assessment will tell you if you are going to be using any chemicals which are hazardous. It should also tell you what to do in the event of spillage. For example, your risk assessment may have identified a particular chemical as being a skin irritant. In which case you need to handle the chemical carefully so as not to spill any on your skin or get it in your eyes. Take care when mopping up spills. Always read the label on the bottle to see if any particular precautions need to be taken and if it is safe to wipe the chemical away. Think about how you are going to dispose of the solutions at the end of your investigation. If your investigation involves heating up liquids decide whether you need to use a large electronic water bath, rather than set up a tripod over a Bunsen on the bench.

2.3 Organised work

The best way to start an investigation is to organise your apparatus. Read your plan so you know exactly what you are going to do. Using your list of apparatus, check that everything you need is gathered together on the bench. Push the items you will not need at the start towards the back of the bench. As part of your preparation, label all the tubes and containers. Think about how you are going to record your data and then prepare some sheets, so you are ready to write down your measurements and observations. Make sure you give yourself time to carry out the investigation. Students often become careless when they are rushing to complete something. Take particular care if your investigation involves the use of live animals. These animals need to be handled as little as possible otherwise they will become stressed and this could affect the results. Make sure you keep them in an appropriate container or cage, placed in a quiet, dark place. Don't allow them to get too hot or dry.

2.4 Making measurements

Part of the implementation assessment will check that you can use the apparatus competently. You should be able to demonstrate a wide range of manipulative techniques with a high degree of skill. These include the skilful use of laboratory equipment and the accurate measurement of volumes, temperatures and times. All the measurements should be taken with care in order to achieve a high degree of precision.

You will be expected to be familiar with the use of measuring cylinders, graduated pipettes and syringes to measure out precise volumes of liquids. Graduated pipettes can be tricky to handle, so familiarise yourself with the correct method of using them. Firstly, you must always use a pipette filler to draw the liquid into the pipette. Never suck the liquid into the pipette. Secondly, you need to check the arrangement of the scale. For example, on a 10 cm^3 pipette, some have the zero level near the tip, whereas others have the opposite arrangement with the zero level near the top. Some pipettes deliver the full volume when the liquid is allowed to drain out, leaving a tiny volume in the tip. Others, called blowout pipettes, deliver the correct volume only when all the liquid has been forced out. Micropipettes are used to measure very small volumes of liquids. Again, you need to know how to set the correct volumes on the micropipette.

When taking readings of levels of liquids in a measuring cylinder, burette or pipette, you must line your eye up with the level of the liquid, and read the volume against the bottom edge of the meniscus.

You may think that making a measurement with a ruler is easy, but there is a source of error known as parallax error. This error occurs when the ruler is not at the same level as the object to be measured, for example, a scale positioned below a capillary tube. If you vary the position of your head relative to the ruler or scale when you make the measurement, you may get a slightly different reading. There are ways of minimising this error. For example, try and position the scale as close as possible to the object to be measured and view the scale so that your line of sight is perpendicular to it.

Many investigations involve the use of stop watches. Take care when reading the stop watch. Many candidates fail to notice that some stop watches record minutes and seconds. The time will be displayed as minutes:seconds, for example 20:15. This reads 20 minutes and 15 seconds, not 20.15 minutes. Although it is acceptable to use minutes and even hours, these units are not strictly SI (see below). The SI unit for time is seconds. So you may want to convert your readings to seconds, especially if the periods of time are less than 10 minutes.

There are some other simple precautions you can take to minimise other errors in your measurements.

- Always stir the fluid in a beaker or flask before taking a thermometer reading.
- Make sure the pan of your balance is clean.
- Check that the balance has been calibrated recently.
- Check that the reading on the balance is at zero before you take any measurements.
- Zero your stop clock or stop watch before starting.
- Always try to use a good quality ruler without damaged ends.

All of your readings should be noted on your record sheets. In addition, you need to make notes of anything unusual which occurred and/or any problems that you encountered.

2.5 Use of SI units

When you make quantitative observations, you are expected to use the appropriate units. The system in use is called SI units (Système International d'Unités). Some of the more common SI units that you will encounter are given in Tables 2.1 and 2.2. It is important that as well as using SI units, you also use them correctly. Table 2.3 gives some of the rules that are associated with the use of SI units.

Table 2.1 *Some common SI units and their symbols*

Name	Unit	Symbol
Mass	kilogram	kg
Length	metre	m
Time	second	s
Area	square metre	m^2

table continued ➢

Name	Unit	Symbol
Volume	cubic decimetre	dm^3
Concentration	moles per cubic decimetre	$mol\ dm^{-3}$
Pressure	pascal	Pa
Energy	joule	J

Table 2.2 *SI units concerned with length*

Unit name	Multiple or fraction of a metre	Symbol
kilometre	10^3	km
metre		m
centimetre	10^{-2}	cm
millimetre	10^{-3}	mm
micrometre	10^{-6}	µm
nanometre	10^{-9}	nm

Table 2.3 *Some rules concerning the correct use of SI units*

Rule	Correct use	Incorrect use
Abbreviations such as sec, cc, or mps should be avoided. Only use standard unit symbols, prefix symbols, unit names, and prefix names	s or second cm^3 or cubic centimetre m/s or metre per second	sec cc mps
Unit symbols are unaltered in the plural	25 cm	25 cms
Unit symbols are not followed by a full stop unless at the end of a sentence	The length of the leaf is 15 cm. The leaf is 15 cm long.	The leaf is 15 cm. long.
A space is used to show the multiplication of units. A solidus or slash is used to show the division of units	The speed of sound is about 344 $m\ s^{-1}$ (metres per second); the decay rate of Cs-113 is about 21 ms^{-1} (reciprocal milliseconds) m/s $m\ s^{-2}$	The speed of sound is about 344 ms^{-1} (reciprocal milliseconds); the decay rate of Cs-113 is about 21 $m\ s^{-1}$ (metres per second) m ÷ s m/s/s

table continued ➢

Rule	Correct use	Incorrect use
Information should not be mixed with unit symbols or names	The biomass of grass was 544 g/m^2.	544 g of grass / m^2
It must be clear to which unit symbol a numerical number belongs	21 cm × 39 cm 20°C to 30°C or (20 to 30)°C 101 g ± 3 g or (101 ± 3) g 74 % ± 6 % or (74 ± 6) %	21 × 39 cm 20°C – 30°C or 20 to 30°C 101 ± 3 g 74 ± 6 %
Unit symbols and unit names must not be mixed	kg/m^3 kg m^{-3} kilogram per cubic metre	kilogram/m^3 kg/cubic metre kilogram/cubic metre kg per m^3 kilogram per metre^3
The word 'weight' is often used interchangeably with the term 'mass'. In science, weight is a force and the SI unit is the newton. The SI unit for mass is the kilogram	The mass of grass is 100 g	The weight of grass is 100 g
The term molarity, with the symbol M, is no longer used. Instead the concentration should be expressed as moles per cubic decimetre or cubic metre	0.2 mol dm $^{-3}$	0.2M

The second (s) is the preferred unit of time, but often it is more convenient to work in minutes, hours, days or years. These can be abbreviated to min or h. It is usual for day and year to be written in full. Often there is confusion between cm^3 and ml. In the past, litre and millilitre were widely used. The litre is still an acceptable unit of volume but its symbol, l, is confusing and can be mistaken for the number 1 (one). Therefore, it is preferable to use decimetre or centimetre cubed. The units of energy are joule (J) and kilojoule (kJ). Do not be tempted to use the calorie as your unit of energy. Unfortunately, the calorie remains a popular unit and most food labels show energy content in both calories and joules.

2.6 Data logging

Data logging is a phrase that describes the use of electronic equipment to record experimental changes. A wide range of sensors can be used, for example pressure, temperature, pH, light and oxygen. The sensor detects a change and produces a voltage that is proportional to that change. The electrical signal is either stored in a portable unit for downloading later or sent back directly to a computer. The data can then be organised and analysed using suitable programs. For example, in an investigation to record the effect of amylase on starch, iodine is added to the solution. Iodine reacts with starch to produce a deep blue-black colour. In this investigation, the rate at which the starch is hydrolysed by the amylase can be monitored using a light sensor. A bright light source is placed to one side of the container and the probe placed on the opposite side. The light shines through the solution and into the probe. As the reaction proceeds, the starch is broken down and the colour becomes less intense. The changes in the light readings are recorded over a period of time. The software for the data logger allows you to set the time intervals for the readings and the number of readings which are taken. At the end, the data is saved to a disk or file on a computer. Data loggers are particularly useful for recording changes at regular intervals over a long period of time. Some of the more advanced data logger units have built-in clocks and calendars, so you can set the system to start recording data when you are not in the laboratory or if you are conducting an ecological investigation and you need remote recording of data over a 24-hour period.

2.7 Making up solutions

Students will usually be expected to be able to make up their own solutions. For example, you specify in your plan that you need a 5% solution of glucose. Take 5 g for glucose and place it in a volumetric cylinder. First, dissolve the glucose in 10 cm^3 of distilled water, then add enough distilled water to bring the total volume to 100 cm^3. This solution is referred to as a 5 per cent weight for volume and it is written as 5% w/v. A 5% solution is not the same as 5% concentration. You need to know the molecular mass of a substance in order to know the concentration. It is generally acceptable to use percentages to describe solutions when you do not know the molecular mass of the substance. For example, you would not be expected to be able to work out the molecular mass of an enzyme, but you could look up the molecular mass of chemicals such as hydrochloric acid and describe the solution in terms of concentration, mol dm^{-3}.

Often you will be provided with a stock solution, which you have to dilute in order to produce the desired solution. Carrying out a serial dilution is a good way of showing your manipulative skills as it involves a number of careful measurements. You need to set up a row of test tubes in a holder on the bench. Add 10 cm^3 of the stock solution of known concentration into the first test tube. Using a graduated pipette, add 5 cm^3 of distilled water to each of the other test tubes. Take 5 cm^3 of the stock solution and add it to the second test tube, bringing its volume up to 5 cm^3. Then take 5 cm^3 of the solution from the second test tube and add it to the third test tube. Repeat this for each of your test tubes. Each time you add the solution to the distilled water you are diluting it by half. For example, if the stock solution was a 10% w/v solution of an enzyme, the second tube would contain 5% w/v solution.

2.8 Working with micro-organisms

If you are carrying out an investigation that involves the culturing of micro-organisms such as bacteria or fungi, there are certain safety procedures that you will have to follow, such as aseptic techniques. The micro-organisms have to be handled correctly to prevent the contamination of your experimental cultures by micro-organisms from external sources and to prevent the contamination of yourself or of the laboratory with any micro-organisms.

Some of the basic rules are as follows.

- When working with micro-organisms, it is important not to eat, drink, lick, chew pens etc. while in the laboratory.
- Any cuts or broken skin must be covered with a plaster.
- Disinfect the bench before you start.
- Loops are used to transfer micro-organisms and these should be sterilised by heating the loop in a flame until the wire is red hot.
- Discarded glassware, loops, etc. should be placed in a container of disinfectant.
- When opening a bottle containing a sterilised solution or a cultures of micro-organisms, prevent cross contamination by sterilising the neck of the bottle. This is done by holding the bottle in one hand and removing the plug or lid with the tips of the fingers of the same hand. Pass the neck of the bottle through a flame. Use a sterilised inoculation loop to remove some of the contents and then stopper the bottle. Do not contaminate the bottle by placing the plug or lid on the bench.
- Any Petri dishes containing cultured micro-organisms should not be opened after incubation in case harmful micro-organisms have been cultured along with the desired ones. The Petri dishes and their contents should be autoclaved before disposal.
- At the end of the session, the bench should be disinfected once again and your hands washed thoroughly.

2.9 Working with microscopes

Your investigation may involve the use of a microscope, in which case your ability to use the microscope could form part of the assessment. You may also need to remind yourself how to make measurements of specimens viewed under the microscope.

Correct use of a compound light microscope

1. Clean the eyepiece and the objective lenses.
2. Rotate the objective lenses so that the low power lens is in line with the main body tube. It should click into the correct position.
3. Place your prepared slide on the stage, making sure that the object on the slide is immediately beneath the objective lens.

4 Looking at the microscope from the side, use the coarse adjustment knob to lower the objective lens so that it is approximately 5 mm above the slide.

5 Looking through the eyepiece, rotate the coarse adjustment knob upwards to bring the slide into focus.

6 Move the slide so that the correct area that you wish to study is in the centre of the field of view.

7 To see the object in greater detail, move the higher power lens objective into place. If you had focussed correctly under low power, you should find that you only need to use the fine adjustment knob to see the object clearly. If you cannot focus under high power, look at the microscope from the side and use the fine adjustment knob to lower the lens so that it is just a millimetre or two above the slide. Then look through the eyepiece and focus on the slide by turning the fine adjustment knob to raise the tube. Always raise the tube upwards as it is very easy to lower the lens onto the slide when using the high power objective.

8 When you have finished, clean the lens and rotate the low power objective into place.

The magnification is calculated by multiplying the magnification of the eyepiece by the magnification of the objective lens. Most eyepiece lenses have a magnification of ×10. There are usually three objective lenses, ×4, ×10, and ×40. This means that, under high power the maximum magnification is ×10 × 40 = ×400.

Measuring objects on a microscope slide

Objects on the slide can be measured using an eyepiece graticule. This is a glass disc that has a tiny scale etched onto its surface. The scale is usually 10 mm long with 100 divisions. The graticule is placed inside the eyepiece so that the scale can be seen in the field of view at the same time as an object.

Before any measurements can be made, the scale has to be calibrated for use with each objective lens. This is achieved by using a stage micrometer. This is a special glass slide that has a series of fine lines etched on its surface. The lines are 1.0 mm, 0.1 mm and 0.01 mm apart. To calibrate, the stage micrometer is moved so that scale on the eyepiece graticule is aligned with the vertical lines on the slide. The length of the eyepiece scale can be measured against the scale of the stage micrometer. This is repeated for each of the lens objectives (see Figure 2.1).

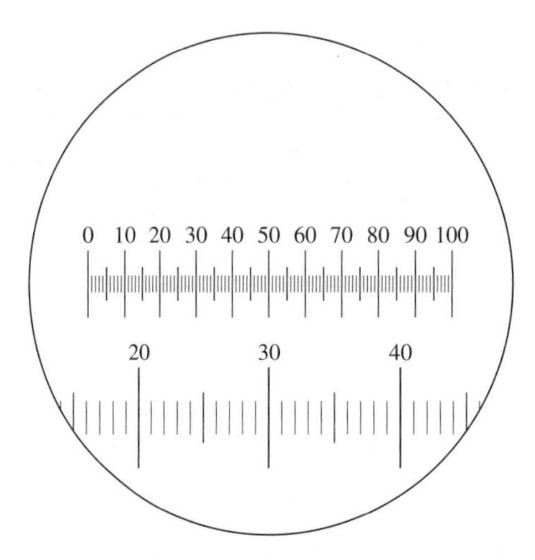

Figure 2.1 *Calibrating an eyepiece graticule using a stage micrometer*

2.10 Recording results

You will be assessed on the way that you recorded your data. Often the quality of your data is a good indication of how accurate and precise you were. The presentation of the data needs to be clear and logical. All your data, both qualitative observations and quantitative values have to be recorded.

Tables of data

Most investigations start with a table of raw data. Take some time to work out how you are going to organise your data in the table. The arrangement needs to be easy for somebody to follow. Often the way in which the measurements are grouped helps when it comes to plotting a chart or graph and carrying out further analysis, such as calculating a rate (see Table 2.4).

Each column in the table should be headed with the quantity together with the SI unit. It is normal to separate the quantity and unit by an oblique or solidus, for example 'Mass/g' or 'Time/s'. Similar labels would be written on a graph. The first column in the table should be the independent variable, that is the variable that you manipulated in the investigation, for example, pH or temperature. The following columns record the values for the dependent variable. These are the measurements that you made. If you have repeated the investigation, then it may be best to group the measurements together, so that you can calculate the mean.

The table in Figure 2.4 has been prepared to record the results from an investigation in which the effects of two types of amylase will be studied over a range of temperatures. The investigation will be repeated three times at each temperature and with each type of amylase. An additional column has been inserted to calculate the mean after the investigation.

Table 2.4 *Sample table showing the layout of data*

Temperature/°C	Time to end point/s							
	Fungal amylase				Bacterial amylase			
	1	2	3	Mean	1	2	3	Mean
15								
20								
25								
30								
35								
40								
45								

Sometimes it is better to draw up a tally chart. For example, you may be measuring the length of leaves to determine the frequency of certain sizes of leaves (Table 2.5).

Table 2.5 *Table showing the tally and frequency of leaves in different size classes*

Leaf sizes/mm	Tally	Frequency
<50	1	1
51–60	111	3
61–70	~~1111~~ ~~1111~~ 111	13
71–80	~~1111~~ 11	7
81–90	11	2

2.11 Anomalous results

An anomalous result is one that seems out of line with all the others that you have obtained in the investigation. You may have repeated the investigation four times and have three replicates that generated similar measurements, but one replicate with very different values. Often an anomalous result does not become apparent until you analyse your data. At that stage there is little that can be done. However, if during the implementation of your investigation you notice that one or two of the measurements look out of line, you do have the opportunity of repeating that bit of the investigation. If the anomalous result was due to experimental error, the repeat should generate values that are in line with the other measurements.

2.12 Writing a report

Writing a report of your method is an essential part of carrying out the investigation. Depending on your awarding body, this may or may not be included in the assessment. Even if it is not, a well-written report will enable the assessor to have a much better idea of what you did during your investigation.

The report should be written in continuous prose rather than as a series of bullet points. The report is always written in the past tense as it is describing the method that you used in the investigation. You have a choice of writing in the first person or third person. In the past, it has been traditional to write in the third person, for example, 'Ten test tubes were taken' or 'The beaker was placed on a tripod and gauze'. However, there has been a recent trend to use the first person, even when writing scientific papers, for example 'I took ten test tubes' or 'I placed a beaker on a tripod and gauze'. A quick glance through some of the most prestigious scientific journals will show that the 'first person' is becoming more usual. Many students find it easier to write in this way. It also shows some personal interest. Remember that you are expected to work on your own, so don't make the mistake of writing 'We took ten test tubes'.

Be precise when your are preparing your report and state exactly what you did. Avoid the use of terms such as approximately, about, fill up, amount, etc. You probably measured out an exact volume of liquid or used measuring scales to weigh out an exact mass of a substance. You probably didn't fill up a test tube. Instead you probably poured in an exact volume. So don't be sloppy in your report writing. It could create a poor impression, when in reality you have been very careful.

Checklist for implementing an investigation

★ Make sure you are familiar with the rules of working safely in the laboratory.
★ Read through your plan before you begin.
★ Check your risk assessment.
★ Make sure you look after any living organisms carefully.
★ Make sure you have all the apparatus you require before you begin.
★ Prepare your record tables for the raw data, with suitable headings and SI units.
★ Organise your apparatus on the bench and label tubes and cylinders, etc.
★ Remind yourself of the correct use of a pipette, burette and any other piece of equipment that you will use during the investigation.
★ If you are using a microscope, make sure you set it up correctly.
★ Measure out all volumes accurately.
★ Make up any solutions that you will require in the investigation.
★ Work through your instructions, making sure that you do not omit any critical stages.
★ Make precise measurements.
★ Make sure you have sufficient time to carry out replicates.
★ Make sure you have plenty of time in which to complete the investigation or that you can complete a part of the investigation within the allocated laboratory time.

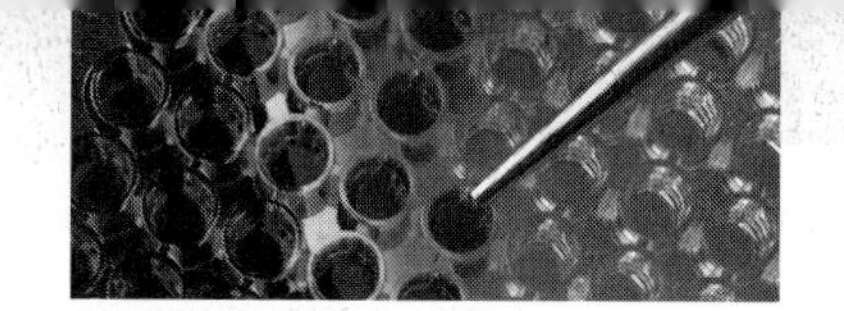

CHAPTER THREE

Analysing

3.1 Key elements of analysis

You should be able to:

- communicate biological information and ideas in appropriate ways, including tabulation, line graphs, histograms, continuous prose, annotated drawings and diagrams
- recognise and comment on trends and patterns in data
- draw valid conclusions by applying biological knowledge and understanding.

Once you have carried out your investigation, you have to present your data in an appropriate format. This could be in the form of summary tables, charts or graphs. In some experiments you may have made qualitative observations in the form of diagrams and drawings. Using your tables and graphs, you need to be able to recognise any patterns or trends shown in the data. From this, you can start formulating your conclusions.

3.2 Summary tables

While you were carrying out your investigation you made measurements and observations, which you recorded in tables of raw data. Now you need to use this data to produce summary tables in which you can show how you have manipulated the data, for example, by calculating the rate of increase or the percentage increase. This helps you to identify any trends in the data.

Table 3.1 shows a set of data obtained in an investigation to determine the time taken for a substrate to disappear at different temperatures. The rate of reaction is the reciprocal of the time taken for the disappearance, that is $1/t$.

Table 3.1 *Time taken for a substrate to disappear at different temperatures*

Temperature/°C	Time taken for substrate to disappear/min	Rate of reaction/min $^{-1}$
10	29.0	0.03
20	16.0	0.06
30	8.0	0.13
40	5.0	0.20
50	9.0	0.11

3.3 Simple maths

Any manipulation of data involves carrying out some simple maths. You are expected to be able to calculate a mean or a percentage change. The mean, or average, is worked out by adding up all the values in a data set and then dividing the total by the number of measurements. The mathematical formula is $x = \Sigma x/n$ where Σ is the sum of, x refers to the individual values and n is the number of values.

For example, $0.5 + 0.6 + 0.3 + 0.3 + 0.4 + 0.7 + 0.3 = 3.1 / 7 = 0.44$

Percentage changes often cause problems, although the calculation is very straightforward. The best way to approach this is to look at the values and work out roughly in your own mind what you think the answer could be. If the second number is more than double the original value then you know you will be expecting more than 100% increase. For example, you have two values, 12 and 15 and you have to work out the percentage increase. The difference between the two values is just 3. To work out the increase, divide the difference between the two values by the original value as follows:

$$3/12 \times 100 = 25\%$$

In an investigation, which involves measurements made over a period of time it is possible to calculate a rate of reaction. This is calculated as $1/t$ where t is the time taken. This is the reciprocal of the value of time.

3.4 Number of decimal places

In general, the number of decimal places in the raw data is a good indication of the accuracy with which the measurements were made. Once you have obtained your raw data, you have to consider how many decimal places to use when processing this data. A general rule of thumb is that the processed data should have the same number of decimal places as the raw data. For example, a number of leaves were measured using a ruler. The ruler allowed the student to measure to the nearest millimetre. The mean value was then calculated. It would be easy for the student to simply copy the value obtained on a calculator. However, calculators usually work out values to several decimal places. But it would be pointless using a mean value with several decimal places as the apparatus did not allow such accuracy. The student would be advised to round up the mean value to the nearest millimetre, to come in line with the rest of the data.

3.5 Plotting graphs and charts

There are a number of ways of presenting a set of data. They include a bar chart, histogram, line graph, scattergram and pie chart.

Bar charts are used when one or more of the variables is not numerical. A non-numerical variable is often referred to as a discrete variable. For example, the number of eggs found in the nest of a particular species of bird, numbers of different types of fruits or blood groups. The bars are of equal width and they do not touch (see Figure 3.1). When the bars are vertical, the chart is sometimes called a column chart.

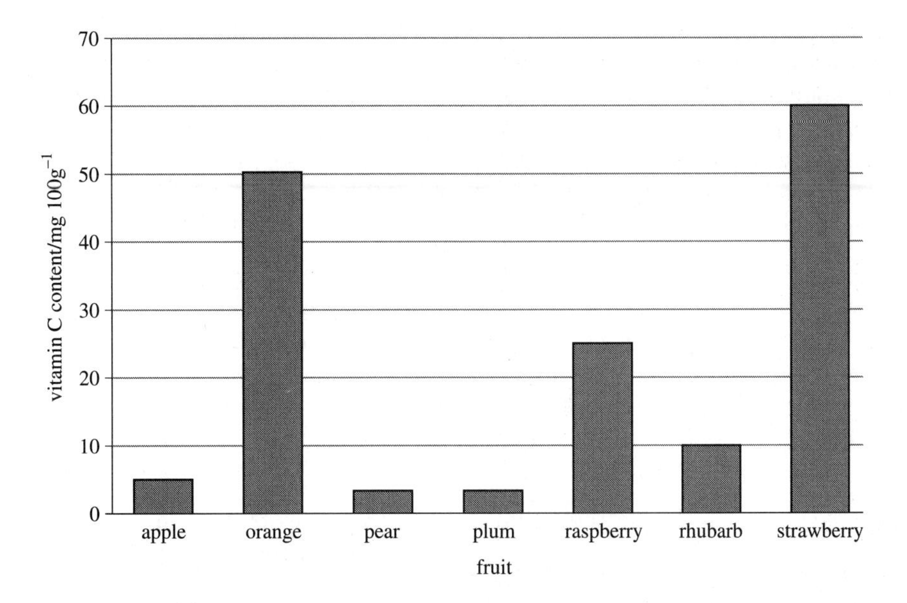

Figure 3.1 *Bar chart showing vitamin C content of different fruit*

A histogram is a special form of bar chart. It is used to plot frequency distribution with continuous data. It would be used, for example, to show the variability of wing length in a fruit fly. The bars are drawn in ascending or descending order and they should touch (see Figure 3.2).

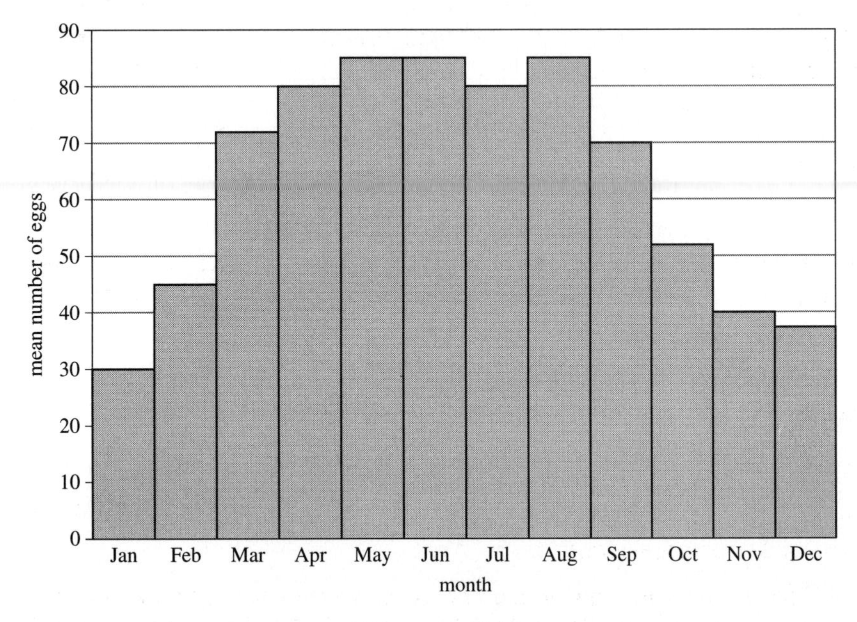

Figure 3.2 *Histogram showing the mean number of eggs laid by a group of chickens over the period of 1 year*

A line graph shows the relationship between two variables, such as rate of reaction against time. The dependent variable is plotted on the vertical or y-axis and the independent variable on the horizontal or x-axis. Time is normally plotted on the x-axis.

All charts and graphs should have a descriptive title. The axes should be drawn the right way round and have appropriate labels with SI units. The scale used to plot the data should be chosen carefully to ensure that the graph fits on the paper and is sufficiently large enough to be read easily. All the points should be plotted accurately. If you are producing a hand-written graph, you may find it better to work with a sharp pencil so that any errors can be corrected. If there is more than one curve, a key needs to be added to the graph.

In a line graph, the points are joined either by a smooth curve or a series of straight lines joining the points. When do you draw a curve and when do you join the points up with a straight line? In most sciences a 'line of best fit' is regarded as the norm and the possibility of joining the points with a straight line is not considered. However, biological data is treated differently. If there is a continuous relationship between the two variables plotted on a graph, a line of best fit should be drawn passing through as many of the points as possible. This option should only be used if there is good reason to think that the intermediate values would fall on that curve. However, in most biological investigations there is no continuous relationship. Instead, the plotted points should be joined by a series of straight lines. Joining points by straight lines indicates that the points in between recorded points are unknown. The values between the points cannot be shown on the graph and cannot be deduced from the readings. Also the way in which they vary between recorded points is also unknown. (See Table 3.2 and Figure 3.3.)

Table 3.2 *Table showing the distance moved by a bubble over a period of time*

Time/min	Distance moved by bubble/mm	
	Leaf A	Leaf B
2	15	5
4	60	7
6	110	15
8	140	22
10	170	25
12	182	28
14	200	33
16	220	37
18	240	44
20	275	56

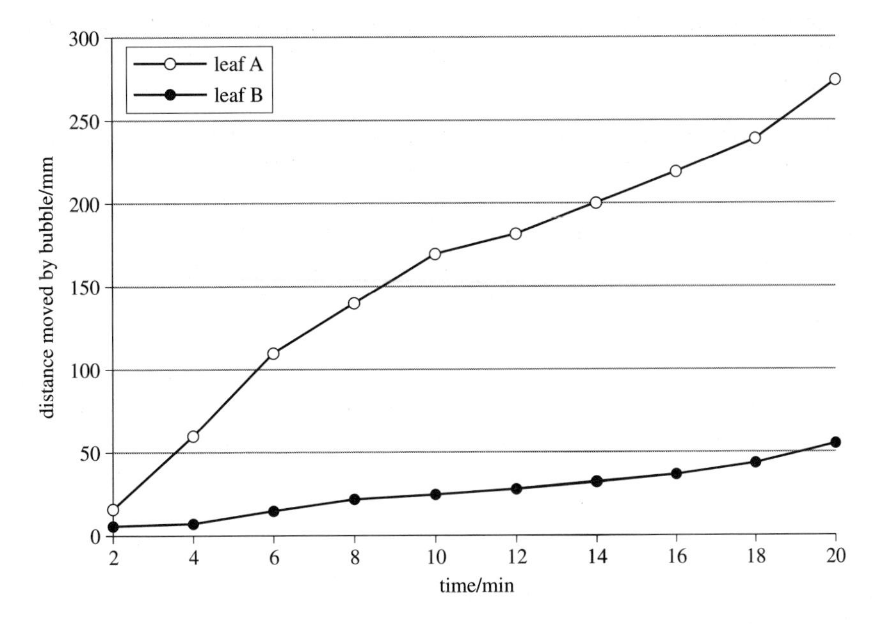

Figure 3.3 *Graph showing the distance moved by bubble over a period of 20 minutes for two different leaves*

A scattergram shows the relationship between two variables. The points on the graph are not joined up but left as single crosses. The way in which the points fall on the graph show the trend.

A pie chart shows the portions of the data of a whole, for example soil is made up of several components – sand, silt, clay – which can be shown on a pie chart. A pie chart could show the breakdown of a population into racial groups. The series of pie charts can be shown in proportion to each other, the overall size of the chart being proportion to the whole sample size. This is usually used on maps to show population size and the breakdown of the population into racial groups. The sectors within a pie chart are arranged in rank order, with the largest first at 12 o'clock, and arranged clockwise.

3.6 Using software programs

It is quite acceptable to use a spreadsheet program such as Microsoft Excel to produce your graphs and charts. However, you must make sure that you are able to use the program effectively. Many students have no knowledge of how to alter the axes, labels and lines, etc. Try not to get carried away and produce lots of graphs. A few well-chosen graphs will create a much better impression. It's also tempting to go for a more unusual method of display, such as a 3D graph, but these types of graph rarely provide more information.

3.7 Recognising trends and patterns

Once you have produced your graphs and charts, take some time to study them carefully. Look at the overall shapes of the curves on your graphs.

- Are they showing an increase or decrease?
- What type of increase or decrease is shown?
- Is it steady or exponential?
- Does the curve level off or plateau?
- Is there a peak, if so, where is the peak?

Keep your descriptions of the trends and patterns concise. You do not need to refer to every single measurement. If you have several curves plotted on the same graph, try and pick out comparative points. Remember that you are aiming to produce an overview or a summary.

3.8 Variability of the data

It is likely that, within a set of data, the values are quite variable. You could make a simple statement about the minimum and maximum values within a set of data. You may want to consider analysing the variability in a bit more detail. If you calculated the mean value and plotted the means on a graph, you could display the variability on the graph as error bars. These are vertical lines that extend above and below the plotted point on the graph. The simplest error bar is one that shows the maximum and minimum values in a set of data. Another method is to calculate the standard deviation. The method for calculating the standard deviation is given in Chapter 5. The standard deviation can then be plotted on a graph as an error bar. The length of the line reflects the value of the standard deviation. Short error bars indicate a small standard deviation.

3.9 Statistical significance

At A2 level you will be expected to carry out a statistical analysis of your data. You will be assessed on whether the test you have chosen is appropriate to the data being analysed and the hypothesis being tested. You need to set out your calculations clearly and show how you looked up the final values in statistical tables. You also need to indicate what level of confidence you decided to use. It is normal to use at least the 5% confidence level. It is also wise to follow this up with a brief sentence stating clearly whether you can accept or reject your null hypothesis. More information on the various statistical tests, level of confidence and worked examples are provided in Chapter 5.

3.10 Conclusions

Your summary of the trends and patterns and the results of your statistical analysis should help you formulate your conclusion. A good place to start is your hypothesis. Do your results support or reject your hypothesis? Try to keep your conclusion concise. It could start with a simple statement. Then you need to add a paragraph or two to explaining your findings using your biological knowledge. Try to keep to information that is relevant to your results, and not to simply regurgitate information from textbooks. Also, you need to think about the biological significance of your findings.

3.11 Anomalous results

It is during the analysis stage of your investigation that you have to decide on how you are going to deal with any anomalous measurement or observation. For example, do you include the measurement in calculating a mean or plot it on your graph? There is no definite rule about dealing with anomalous measurements. If you have a large set of values and only one or two are very different to the other values, then you are probably justified in ignoring these values and you could omit them from your calculations. The inclusion of these anomalous values in the calculation for a mean or in a statistical test could distort the results and lead you to the wrong conclusion. However, if there appears to be one anomalous value out of a small set of three or four replicate values, it is more difficult to justify ignoring the value. When plotting graphs, anomalous values could be included on a graph, but not actually joined up to the other points. Whatever you decide to do, it is important to comment on this in your report.

Checklist for analysing an investigation

- ★ Have you recorded your results in a table?
- ★ Has your table got a clear title and labelled columns with SI units?
- ★ Have you drawn a graph with a title, scale, labelled axes, accurate plots with a neat line joining them up?
- ★ Have you carried out an appropriate statistical test?
- ★ Have you explained the validity of your result with confidence limits?
- ★ Are your results significant?
- ★ Have you written a clear and concise conclusion?
- ★ Have you referred back to your hypothesis?
- ★ Have your quoted examples from your data to support your conclusion?
- ★ Have you analysed your results using biological knowledge?
- ★ Have you considered anomalous results?

Example: AS level analysis

A student produced Table 3.3 and graph (Figure 3.4) as part of their analysis of their investigation. The investigation studied the effect of pH on the activity of catalase in potato tissue.

Table 3.3 *Table showing the mean volume of oxygen produced at three different levels of pH*

Time/min	pH 5	pH 7	pH 9
0	0	0	0
1	2	5	3
2	5	15	7
3	7	24	11
4	11	33	15
5	15	45	22
6	19	51	29
7	25	62	29
8	27	70	33
9	29	73	39
10	33	79	42
11	35	83	45

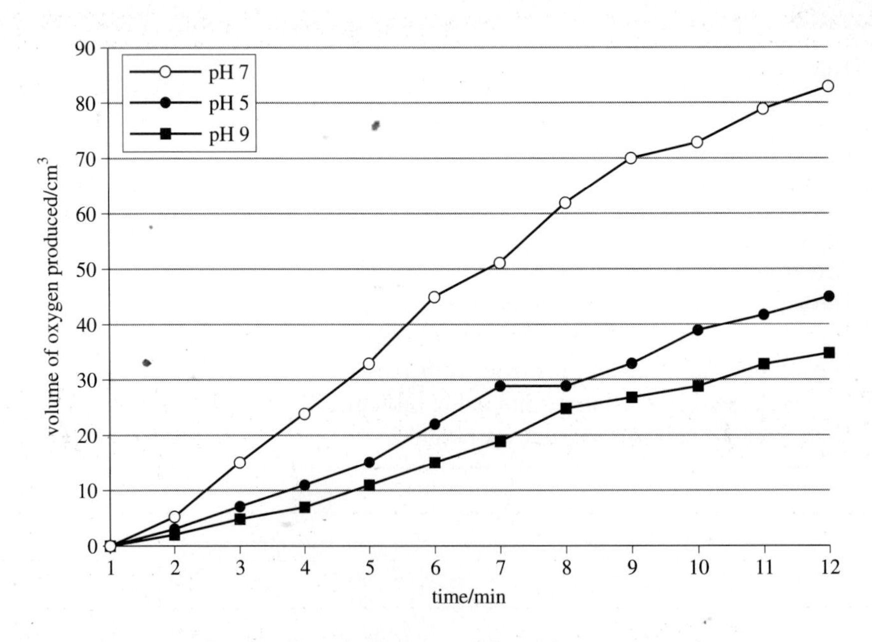

Figure 3.4 *Graph showing the effect of pH on the activity of the enzyme catalase*

Teacher's comments

The table of results could have been much better. There is no indication of how many values were used to obtain the mean values or whether the values have been rounded up or down to the nearest whole number. The investigation looks at oxygen production over a period of time, so the student should have calculated the rate of reaction. Ideally, the student should have produced two tables: the first table with all the raw data and the mean values and a second table with the rate of reaction.

The graph is well presented. It has a clear title, the axes are labelled, a suitable scale has been used, the points are clearly shown and neatly joined up and there is a key. The graph clearly shows the following trends:

- there is an increase in the production of oxygen at all three pH values over the 12 minute period
- the volume of oxygen produced at pH 7 at 12 minutes is approximately double that produced at pH 5 and pH 9
- the increases in oxygen production for pH 5 and pH 9 are similar.

It is possible to conclude that pH does affect the activity of catalase in potato. The activity of this enzyme is lower at more acidic and more alkaline pHs than at neutral pHs. This conclusion could be backed up by including some enzyme theory such as the effect of pH on enzyme structure, particularly on the shape of the active site, the disruption of bonding within the enzyme, denaturation, etc. However, it would not be possible to conclude that pH 7 was the optimum pH for this enzyme. There are only three pH values which is very limiting. Further testing using a much narrower range of pH values around pH 7 would be necessary before this conclusion could be made.

Example: AS level analysis

An investigation was set up to investigate the effect of grazing by caddis flies on the algal population of a stream. The student placed small, unglazed ceramic tiles on the bottom of the stream and observed the colonisation of these tiles by algae and caddis flies over a period of 7 weeks. For each tile, the student measured the percentage cover of algae and the number of caddis flies that were grazing on the tile (see Table 3.4 and Figure 3.5).

Table 3.4 *Table showing the percentage cover of algae and the number of caddis flies over a period of 7 weeks*

Weeks of colonisation	Percentage cover of algae					Number of caddis flies per tile				
	1	2	3	4	Mean	1	2	3	4	Mean
0	0	0	0	0	0	0	0	0	0	0
1	65	49	45	57	54.0	25	21	27	19	23.0
2	80	82	75	80	79.3	30	35	18	15	24.5

table continued ➢

Weeks of colonisation	Percentage cover of algae					Number of caddis flies per tile				
	1	2	3	4	Mean	1	2	3	4	Mean
3	68	61	45	50	56.0	137	120	101	140	124.5
4	27	21	25	32	26.3	103	95	110	99	101.8
5	13	15	20	30	19.5	100	105	97	89	97.8
6	35	30	40	25	32.5	95	97	95	85	93.0
7	80	90	75	95	85.0	50	44	51	15	40.0

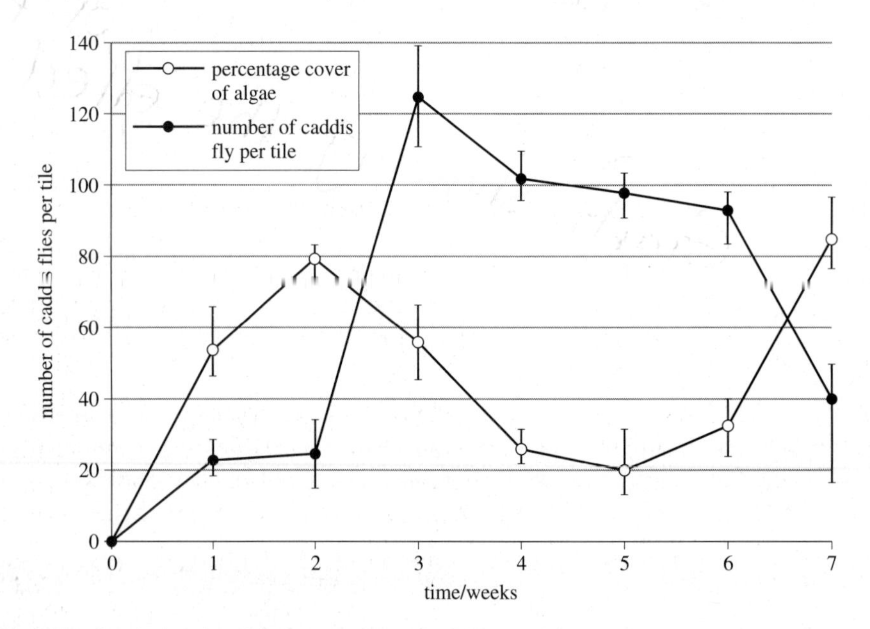

Figure 3.5 *Graph showing the percentage cover of algae and the number of caddis flies on the tiles over a period of 7 weeks*

Analysis

During the first week, the algae started to colonise the tiles. They grew and reproduced and gradually covered the surface. As the algae increased, the number of caddis flies started to increase too. The algae peaked at week 2 with a mean coverage of just under 80%. The caddis flies continued to increase and peaked at week 3 with a mean of 124.5 flies. Both decreased for a few weeks. After week 5 the algal coverage started to increase. The decrease is particularly steep after week 6.

Both of the organisms increased over the first few weeks, but the numbers of the predators, the caddis flies, increased less steeply than the algae. Algae grow and reproduce quickly. They can undergo asexual reproduction and take advantage of fresh surfaces to colonise. The caddis flies were attracted to the tiles because they eat algae. As there was more algae, more caddis flies moved onto the tiles. With three weeks the algae had covered 80% of the tiles and this provided a lot of food for the caddis flies and so their numbers shot up. There were so many caddis flies grazing on the algae that they started eating more than was produced.

The growth and reproduction rate of the algae could not keep up. Once the coverage of algae fell to a low value, the caddis flies moved off in search of food elsewhere. Eventually, the numbers of caddis flies fell back so the algae had a chance of recovering. When this happened, more caddis flies started to move back and the over grazing happened again. The peak of the algae never coincided with that of the caddis flies. One lagged behind the other. This is an example of a predator–prey relationship.

I plotted error bars on my graph. The variability of the sets of data fluctuated, but there was little overlap from week to week.

Teacher's comments

This student has carried out an interesting investigation and produced some good data. She has presented the data in an appropriate table, showing the replicate values and the mean. The mean has been rounded up to one decimal place. This could have been rounded up further to the nearest whole number to come in line with the rest of the data, particularly since some of the data refers to caddis flies, which come in whole numbers!

The student has chosen to draw a line graph. The graph is titled, the axes are labelled and the scale is appropriate. There are two curves, each of which is identified in the key. Maximum and minimum error bars have been added to give an indication of the variability in the data. One minor criticism is that the time axis should have been extended to 8 weeks in order that the last points with their error bars could be seen more clearly.

The student has described in some detail the peaks and troughs in the data and identified a lag between the algal cover and the caddis flies. However, no mention has been made of any anomalies. There is one possible anomaly for week 7. One of the values for the caddis flies has fallen to 15 while the other values are about 50. This has dragged the mean value down considerably and created a very variable set of data, as indicated by the length of the error bar. Even if the student did not think that there were any anomalies, she would have been wise to state that she checked the data for any anomalies. However, the candidate has given a good explanation for the relationship between these two organisms and has used appropriate biological terminology.

Overall, this was a good analysis, although it would be prevented from gaining the highest marks because there is no reference to anomalies.

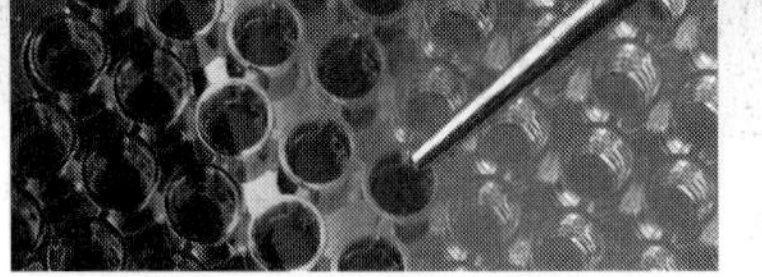

CHAPTER FOUR

Evaluation

4.1 Key elements in evaluation

You should be able to:

- assess the reliability and precision of experimental data and the conclusions drawn from it
- evaluate the techniques used in the experimental activity, recognising their limitations.

Once you have analysed your results, you are ready to complete an evaluation of your investigation. The evaluation looks at the reliability and precision of your results and assesses the limitations of your chosen method.

4.2 Reliability of data

A good set of reliable data would be one in which the results could be repeated. If somebody else were to carry out exactly the same investigation, following the method described, they should be able to obtain similar data.

When you are considering the reliability of your data, you should not write, 'I may not have read the measurements accurately enough' or 'if I had better equipment'. Instead, you need to identify problems with your method which could have meant that data was not reliable. You may have had too few replicates and so the reliability of your investigation could be improved by increasing the number of replicates.

You need to comment on the variability of your data. If your replicate values are close together, this indicates that the data are more reliable than a set of data with widely differing values. You may have calculated the standard deviation on a set of replicates, which gives you a measure of the range of values for one set of data.

4.3 Precision of data

The precision of data refers to the accuracy of the results. Perhaps your method of making measurements was too hit and miss. For example, the end point of a titration was difficult to determine, so you did not make the measurement of the end point at exactly the same point each time. Often, experiments which involve timed measurements based on colour changes can be inaccurate and it would be better to use a colorimeter to improve the reliability. You may have misread burette readings by not always taking the reading from the bottom of the meniscus. There may have been a parallax error when reading off a scale.

It is possible that you could have improved the precision by using equipment that would give you more detailed measurements, for example a precision balance or a vernier caliper rather than a short ruler. Perhaps you were monitoring pH changes using indicator paper when it would be better to use an accurate pH meter.

4.4 Limitations of your method

This part of the evaluation requires you to look carefully at your method and decide if there were any ways in which you could improve the design to get more accurate results. In an investigation involving enzymes and temperature, you may have chosen far too wide a temperature range. This would mean that you were not able to determine the optimum temperature. So, one improvement would be to repeat the investigation, but this time use a much narrower temperature range with smaller intervals between the temperatures.

There may have been sources of error in your procedures. For example, you could have carried out an investigation into catalase, using potato as a source of the enzyme. You diced up the potato before placing it in the hydrogen peroxide. However, unless the diced pieces of potato were exactly the same in number and size, you would have no way of knowing that the surface area of the potato exposed to the hydrogen peroxide was the same in all the treatments.

4.5 Further study

Often an investigation throws up more problems than answers. So as well as indicating how the investigation could be improved, you should add some ideas for further study with a brief outline of how you would design it. The further study should provide some additional evidence or extend the original investigation.

4.6 Anomalous results

Anomalous results have been mentioned in several of the preceding chapters. You should comment on any anomalous results in your evaluation. If your investigation produces a number of anomalous results, you may have to look more carefully at your method or preparation in case an error has crept in and caused these odd results.

Checklist for carrying out an evaluation

- ★ Have you commented on the reliability of your results?
- ★ Have you commented on the variability of the replicate values?
- ★ How could you improve the reliability of your data?
- ★ Have you commented on the precision or accuracy of your measurements?
- ★ Have you commented on any sources of error?
- ★ Have you commented on the limitations of your methods and apparatus?

Checklist for carrying out an evaluation continued

- ★ Have you considered any anomalous results?
- ★ Have you made any proposals for further investigations?

Example: AS level analysis and evaluation

An investigation into the effect of barley straw on algal growth

A student's mother had read that she could stop algae growing in her pond by filling the leg of an old pair of tights with barley straw and floating it in the water. The student planned and carried out an investigation into the effect of barley straw on algal growth.

She collected some pond water with algae and placed it in five 2-litre clear plastic containers. She placed lamps overhead. She measured the number of algae per mm^3 using a haemocytometer. The first container was left as a control. She added the following quantities of finely chopped barley straw to the containers: 25 g, 50 g, 100 g and 150 g. The straw was placed in the toe of a pair of tights with the end tied off before being immersed in the pond water. She took samples of pond water every 3 days for a total of 20 days. Each container of pond water was sampled three times and a mean value was calculated.

Her tabulated data, graphs, analysis and evaluation are given below (see Tables 4.1 and 4.2 and Figures 4.1 and 4.2).

Table 4.1 The number of algal cells observed using a haemocytometer from day 0 to day 20

Mass of barley straw/g	Mean number of algal cells counted in haemocytometer/$mm^3 \times 10^3$				
	Day 0	Day 3	Day 6	Day 9	Day 12
0	55	75	157	250	344
50	51	82	143	209	337
100	45	77	126	131	169
150	52	64	75	88	95
200	54	61	69	85	60

***Table 4.2** The rate of increase in the number of algal cells from day 0 to day 20*

Mass of barley straw/g	Rate of increase of algal cells/number per day × 10^3				
	Day 0	Day 3	Day 6	Day 9	Day 12
0	0	6.7	27.3	31	31.3
50	0	7.0	20.3	22.0	42.7
100	0	10.7	16.3	1.7	12.7
150	0	4.0	3.7	4.3	2.3
200	0	2.3	2.7	5.3	−8.3

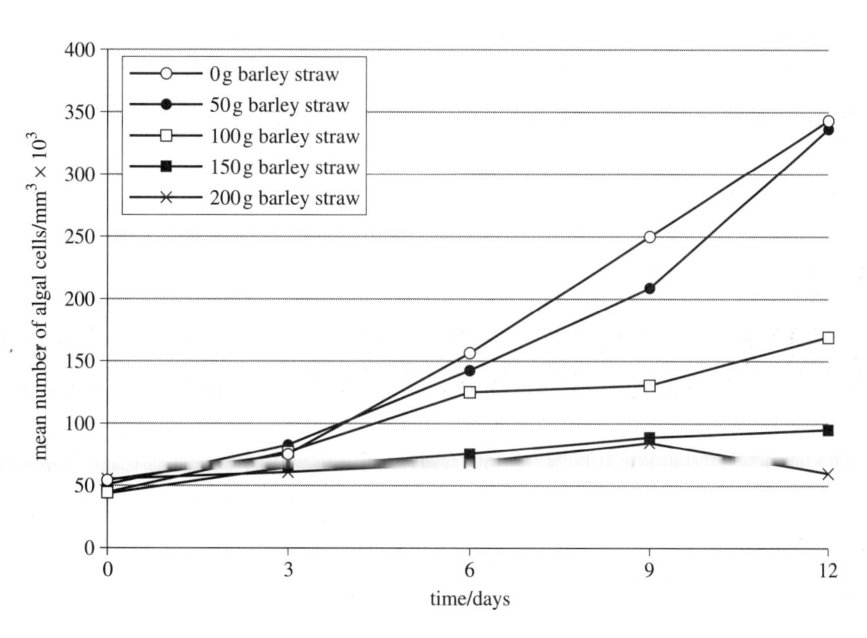

***Figure 4.1** Graph showing the increase in the number of algal cells over a period of 12 days for five different barley straw treatments*

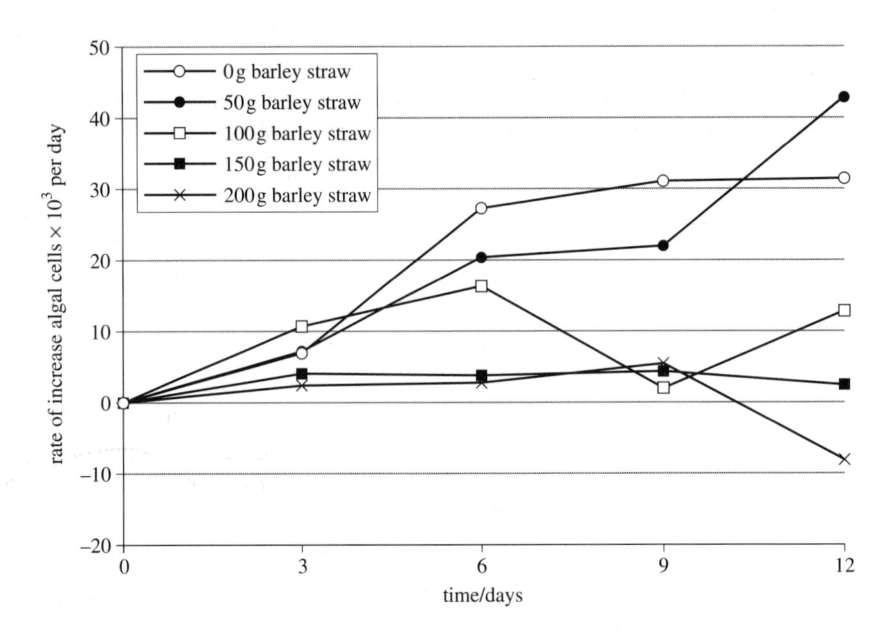

Figure 4.2 *Graph showing the rate of increase of algal cells over a period of 12 days for five different barley straw treatments*

Hypothesis

Barley straw releases a chemical into the water that kills algae. The more barley straw added to the water, the fewer the numbers of algal cells.

Analysis

The results support my hypothesis. As the quantity of barley straw increases, the number of algal cells in the water decreases. The control contains no barley straw. During the 20 days the number of algal cells in the control water increased from 55 to 244×10^3. This is a 525% increase. The results for 50 g of barley straw are similar to that of the control. There was a 560% increase. This is slightly higher than the control. This seems to suggest that 50 g of barley straw was not enough to have an effect on the algae. When 100 g of barley straw was added, the number of cells still increased, but the increase in the number of algal cells was less. Overall, there was a 275% increase. However, the result for day 9 was unexpectedly low. This is probably an anomalous result. There is little difference between 150 and 200 g of barley straw. Both treatments slowed down algal growth. By the end of the investigation the algal cells in the 200 g treatment had started to die.

My results indicate that the barley straw contains a substance that moves into the pond water. This substance slows down the growth rate of the algae. It could be an inhibitor of some sort. However, there has to be quite a lot of the substance in order to stop growth all together and kill the algae. It is possible that this substance inhibits enzyme action. This would slow down the growth of the algae and possibly stop them reproducing.

My results suggest that this treatment could be used in garden ponds to stop algae from turning the pond water green. However, it would be a bit unattractive to have a large quantity of barley straw floating in the pond. It could also be used to treat rivers which are suffering from algal blooms. It is possible to buy whole bales of barley straw and they would float in the river water.

Further study

It would be useful to know more about the inhibitor and to identify it. I could repeat my investigation using larger quantities of barley straw to determine if there is an optimum mass which should be added. I would also like to know the effects of having too much inhibitor in the water. It could kill other aquatic organisms.

Limitations

There was not too much variation between the three samples taken to calculate the mean. However, there was one odd result. The result for day 9 for 100 g of barley straw was amongst the lowest. It is possible that the samples I took were not representative of the whole water.

The number of algal cells were counted using a haemocytometer. This was a relatively accurate method of counting the cells. However, the volume of pond water sampled was very small so it may not have been representative, even though I took three samples. Towards the end of the investigation there were a lot of algal cells and in some of the samples it was difficult to count the individual cells in the haemocytometer as they were clumped up together. Another method would have been to use a colorimeter. This would have measured the amount of light passing through the sample.

The barley straw was chopped up and placed in the toes of old tights. The chopping process was not accurate and there were a mix of lengths of straw. This may have meant that the surface area of the barley straw in contact with the water could have varied. The larger masses of barley straw had to be squashed into the tights, whereas the small amounts of straw were less compacted. This too could have affected the spread of the inhibitor from the straw.

Teacher's comments

The data has been carefully displayed in appropriate tables and on graphs. The student has calculated the mean number of algal cells and their mean rate of increase or decrease. These values were then plotted on graphs. The graphs have correct titles, labels and scales. The points are joined up correctly and accurately. SI units have been used throughout.

The student has identified the overall trend and there are suitable comparisons with good use of the figures. The anomalous result has been commented upon. There is a reasonable attempt to explain the results and to link them to biological knowledge. There is a brief reference to enzymes and inhibitors. The student could have provided more information on inhibitors and how they work.

In her evaluation, the student has commented on the lack of variability in her data and identified a possible anomaly. However, she has not really commented on the reliability and precision of her data. There is a brief mention that the values obtained with the haemocytometer were accurate. However, she has assessed the limitations of her method. She has identified possible problems with the sampling of the pond water and the difficulty of counting the individual cells when there were large numbers in the haemocytometer. She has proposed a further investigation which would examine the nature of the inhibitor in greater depth.

Overall, this was an average to above average analysis and evaluation. All the main criteria have been considered. However, this is not a top scoring assessment as there are some areas of weakness which could be improved. For example, in the analysis, more detailed explanation is needed and in the evaluation more emphasis is needed on the variability of the data.

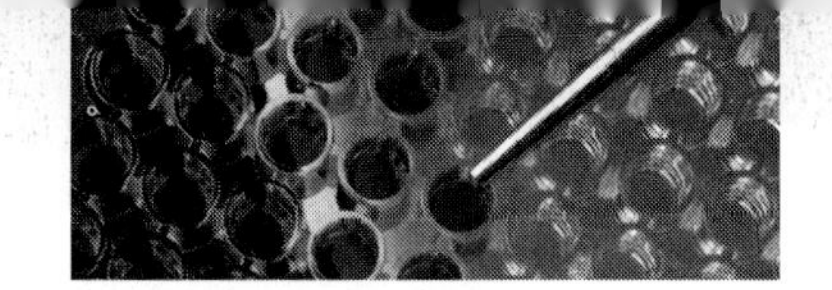

CHAPTER FIVE

Statistics and IT

One of the important differences in coursework at AS and A2 level is the requirement for a statistical test to be included in the design of the investigation. The statistical test is carried out to prove whether the results are significantly different or not. It cannot be stressed enough that one of the first decisions to be made when planning an investigation is the type of statistical test that will be used. The choice of test will in many cases determine the design of the investigation and the number of replicates.

There are a handful of tests which need to be considered. Often, students are most familiar with the Student's *t*-test and the Chi-squared. However, tests such as Mann–Whitney *U* test, the Wilcoxin matched pairs test and the Spearman rank correlation are far more useful.

5.1 Variable data

Biological investigations tend to produce variable measurements. No one measurement can be considered to be typical, for example measuring the height of a single student in a class would tell you very little about the heights of people of that age. Instead, it is necessary to measure a larger sample. Once this data has been obtained it has to be analysed in some way. Most students are tempted to calculate the mean value of a set of measurements. The mean value is fine if the data are normally distributed. If the data do not fit a normal distribution, but are skewed instead, then calculating the median value is more useful (see Figure 5.1).

For example, a set of data consists of the following values: 1, 5, 7, 66, 34, 20, 12, 7, 8, 2.

The mean is 162 / 10 = 16.2 while the median value is 7.5.

It is easy to determine whether the data has a normal distribution by producing a simple tally chart (see Table 5.1). The values are divided up into 8–10 size classes and the number of occurrences in each class is counted up. It can be difficult to decide whether data in a small set is normally distributed, so it is probably safer to assume that it is not.

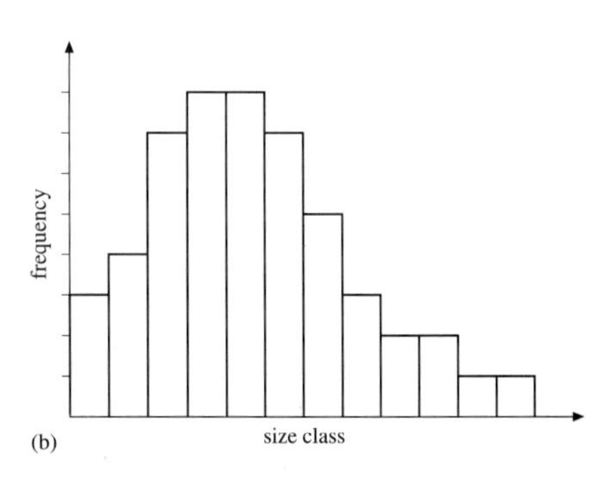

Figure 5.1 ***(a)*** *Normal distribution and* ***(b)*** *Non-normal or skewed distribution*

six replicates of all the measurements. If you have 12 or more samples, you have just enough data to check if your data are normally distributed or not. Ideally, your method of sampling (if applicable to your design) would be random to ensure there is no bias. If you have more than 15 replicates and a normal distribution, you can summarise your data using a mean value and test the difference using the Student's *t*-test. If you have less than 15 replicates and a skewed distribution, you can use the Mann–Whitney *U* test. Finally, if you have matched pairs, you can use the Wilcoxin matched pairs test. In this test a measurement from one set of data is matched to a measurement in a second set of data. This would be appropriate for an investigation in which you were comparing the pulse rate for a person before and after exercise, for example.

Many investigations look for an association between two uncontrolled variables. For example, to test whether there is a relationship between the number of freshwater shrimps and the velocity of water in a stream, or the relationship between light intensity and the percentage cover of a certain plants along a woodland transect, the best test for carrying out a correlation is the Spearman Rank Correlation test. In this test you will need at least 12 pairs of measurements. You may want to be able to control one variable while you measure the effect of changes of this variable on a second dependent variable. For example, you are measuring the change in the rate of oxygen production at different temperatures. The test to carry out is called a correlation test. If you want to determine the line of best fit, then carry out a regression analysis.

A few investigations involve a biological theory, which predicts a set of expected values. The set of expected or predicted results are compared against a set of observed values, for example genetics and behavioural investigations and some ecological investigations which test presence/absence associations. These investigations should use the χ^2 (Chi-squared) test.

5.4 Null hypothesis

Whichever statistical test you decide to use, you will have to produce a hypothesis. A hypothesis is simply an idea or prediction – what you think will happen. The aim of your investigation is to produce evidence that will prove or disprove the hypothesis. Proving that a hypothesis is correct is surprisingly difficult, but it is much easier to find evidence that will disprove it. Therefore, it is more usual to produce a negative, or null, hypothesis. For example, there is no difference in the number of oak trees found in two different woods. If the data prove that the null hypothesis is not correct, then there must be a difference. The null hypothesis can be rejected.

5.5 The Students *t*-test

You should use the Students *t*-test if you want to compare two sets of normally distributed data. The test looks at the degree of overlap between two sets of data. The more overlap that is present, the less likely it is that the data are different.

The value of t is large when there is a large difference between the means of the two samples and the data are clustered tightly around these means. However, the value of t will be small if the difference between the two means is small and the data are widely spread. So, the larger the value of t, the more certain it is that the two sets of data are different.

The value of t is checked against a table of critical values. If t is greater than, or equal to, the critical value, then you can accept that there is a statistical difference between the two sets of data. The critical value takes into account the reliability of the data. A large sample size would be more reliable than a small sample. So, the critical value depends on the number of measurements that were taken. In order to look up the correct critical value, the degrees of freedom (DF) have to be calculated.

$$\text{DF} = n_1 + n_2 - 2$$

There are three levels of significance. It is usual to use the 5% significance level. This means that there is a 1 in 20 chance of obtaining a value of t equal to the critical value, purely due to random variation. If the value of t is greater than the critical value, then there is a significant difference between the two sets of data.

Worked example

A student investigated the variation in the length of mussel shells on two locations on a rocky shore. A t-test was carried out to see if there was any significant difference between the two samples (see Table 5.3).

Null hypothesis: there was no difference in the length of the shell.

Table 5.3

	Shell length for group A/mm		Shell length for group B/mm	
n (number)	x_1	x_1^2	x_2	x_2^2
1	46	2116	23	529
2	50	2500	28	784
3	45	2025	41	1681
4	44	1936	31	961
5	63	3969	26	676
6	57	3249	33	1089
7	65	4225	35	1225
8	73	5329	21	441
9	55	3025	38	1444
10	79	6241	30	900
11	62	3844	36	1296
12	59	3481	38	1444
13	71	5041	45	2025
14	68	4624	28	784

table continued ➢

Worked example *continued*

	Shell length for group A/mm		Shell length for group B/mm	
15	77	5929	42	1764
Total for each column	Σx_1 914	Σx_1^2 57534	Σx_2 495	Σx_2^2 17043
Calculate mean by dividing by n, where $n = 15$	$\Sigma x_1/n$ 61	$\Sigma x_1^2/n$ 3835.6	$\Sigma x_2/n$ 33	$\Sigma x_2^2/n$ 1136.2
Mean squared	x^2 3721	–	x^2 1089	–
Calculate the variance (s^2) $\Sigma x^2/n - x^2$	3835.6−3721 $s_1^2 = 114.6$		1136.2−1089 $s_2^2 = 47.2$	

Σ = sum of.

To calculate the value of t:

$$61-33/\sqrt{114.6/15 + 47.2/15} = 8.53$$

This value is checked against the critical values for the t-test. The degrees of freedom for this investigation are calculated as follows:

$$\begin{aligned} DF &= n_1 + n_2 - 2 \\ &= 15 + 15 - 2 = 28 \end{aligned}$$

The critical value at $p = 0.05$ is 2.05. Since the value of t is greater than this value, the null hypothesis can be rejected.

5.6 Mann–Whitney *U* test

This test is probably the most useful of all the statistical tests and it is relatively straightforward. It should be used when the data from the investigation are not normally distributed, but skewed. This means that the mean value is not very useful. Instead the median value is used (the middle value in the range). The Mann–Whitney *U* test compares the median of two sets of data. A quick way to determine whether it is worth carrying out this test is to calculate the median value of each set of data. If the median is the same in both, there is little point in continuing with the analysis.

The stages in this test are as follows.

1 Produce a table that has three columns and write your raw values for one set of data in the first column.

2 In the middle column, rearrange the data into order, from the lowest value to the highest.

3 In the third column, write in the ranks, giving the lowest value the lowest rank. If there are two values the same, they share the rank (R).

4 Add the ranks up to get the sum of the ranks.

5 Repeat this with the other set of data.

6 Use the following formula to calculate the U values.

$$U_1 = n_1 \times n_2 + 1/2n_2\,(n_2 + 1) - \Sigma R_2$$

$$U_2 = n_1 \times n_2 + 1/2n_1\,(n_1 + 1) - \Sigma R_1$$

The smallest U values are obtained when there is no overlap between the two sets of data. Using the smaller of the two values, look up the critical value at 5% in the tables of U statistics. If the U value is equal to, or less than, that in the table, the null hypothesis can be rejected. The difference between the two samples is significant at the 5% level.

Worked example

A student investigated the number of mayfly nymphs in two different habitats, a shallow pool and a deep pool in the same river. He took samples for 1 minute at ten randomly selected sites and obtained the results shown in Table 5.4.

Table 5.4

Sample	1	2	3	4	5	6	7	8	9	10	Median
Shallow	10	9	15	15	11	15	20	16	21	9	15
Deep	2	3	5	3	7	4	11	12	6	8	5.5

First, arrange the data in order and then in rank order. The lowest value is given the lowest rank, the highest value the highest rank. Note how the same values are given the same averaged rank. For example, there are two values at 9. They would have occupied ranks 9 and 10. So, they are given the same shared rank calculated by adding up the two ranks and dividing by 2, for example 9 + 10 = 19/2 = 9.5. The next rank to allocate is 11. Where three values are the same, you should add up the value of the three ranks and divide by three (Table 5.5).

Table 5.5

Shallow data	Ranked order	Rank (R_1)	Deep data	Ranked order	Rank (R_2)
10	9	9.5	2	2	1
9	9	9.5	3	3	2.5
15	10	11	5	3	2.5
15	11	12.5	3	4	4
11	15	16	7	5	5
15	15	16	4	6	6

table continued ➤

Worked example continued

Shallow data	Ranked order	Rank (R_1)	Deep data	Ranked order	Rank (R_2)
20	15	16	11	7	7
16	16	18	12	8	8
21	20	19	6	11	12.5
9	21	20	8	12	14

Sum the ranks for each set of data:

$\Sigma R_1 = 9.5 + 9.5 + 11 + 12.5 + 16 + 16 + 16 + 18 + 19 + 20 = 147.5$

$\Sigma R_2 = 1 + 2.5 + 2.5 + 4 + 5 + 6 + 7 + 8 + 12.5 + 14 = 62.5$

Calculate U_1 and U_2 values using the following formula:

$U_1 = n_1 \times n_2 + 1/2n_2\,(n_2 + 1) - \Sigma R_2$

$= (10 \times 10) + 5 \times (10 + 1) - 62.5 = 92.5$

$U_2 = n_1 \times n_2 + 1/2n_1\,(n_1 + 1) - \Sigma R_1$

$= (10 \times 10) + 5 \times (10 + 1) - 147.5 = 7.5$

where n_1 and n_2 are the number of samples.

Using the smaller of the two values (in this case, U_2 at 7.5), look up the value in the tables of U statistics. The critical value at 5% is 23. Since the smallest U value is less than that in the table, the difference between the two samples of mayflies was significant at the 5% level.

5.7 Wilcoxon matched pairs test

This test compares individual values in one set of data with individual values in a second set. In order that the data can be considered as matched pairs, there must be a single unique way in which the values can be linked together. A good example is an investigation in monitoring the pulse rate in a group of athletes before and after exercise. The pulse rate before exercise is matched against the pulse rate after exercise for the same person. Another suitable investigation could be one in which a sample of yeast was split into two, half grown on one type of sugar, and the other grown on a different sugar. This test requires at least six sets of replicates. With less than ten replicates, the data need to be near perfect with the differences being in the same direction, i.e. all positive differences or all negative. Ideally, you should have between 10 and 15 replicates. The test can deal with up to 30 replicates, but the data is more difficult to handle.

The calculation is carried out as follows.

1 Calculate the differences between each pair of measurements.

2 Ignore the signs and rank the differences. These are referred to as non-zero

differences. Any pairs that have no difference in value are excluded from the test. The lowest rank is given to the smallest difference.

3 Now look at the signs and calculate the sum of the positive ranks and the sum of the negative ranks:

$$\Sigma R_{-ve} = x \text{ and } \Sigma R_{+ve} = y$$

4 Compare the smallest of the ΣR values against the critical value for the number of non-zero differences. If the smallest ΣR value is less than or equal to the critical value, then the null hypothesis can be rejected.

Worked example

A student carried out an investigation into the change in pulse rate before and after exercise. The student asked 12 students of the same sex and age to measure their pulse while they were sitting down. Then they ran 100 m and measured their pulse rate again. The results are shown in Table 5.6.

Table 5.6

Person	A	B	C	D	E	F	G	H	I	J
Pulse before	63	64	65	72	72	69	79	67	71	75
Pulse after	109	165	123	176	112	180	156	145	171	138
Difference	46	101	58	104	40	111	77	78	100	63
Rank	2	8	3	9	1	10	5	6	7	4
+ / −	+	+	+	+	+	+	+	+	+	+

$$\Sigma R_{-ve} = 0$$

$$\Sigma R_{+ve} = 55$$

The lowest value of ΣR is 0. The critical value for ten non-zero differences at 5% is 8. The ΣR value of 0 is less than the critical value. This shows that there is a significant difference between the pulse rate before and after exercise.

5.8 Correlation

A correlation is a statistical method that answers the question: are these two variables associated? In other words, if one variable changes, the other changes too. For example, a relationship between the soil pH or light intensity and the percentage cover of a particular plant species.

The Spearman rank correlation test determines if the relationship between two variables is significant. The test involves the calculation of a correlation coefficient, r. It can range from 0 when there is no relationship, to +1 when there is a perfect positive correlation and −1 when there is a perfect negative relationship. A positive correlation indicates that as one variable increases, the other increases too. A negative correlation shows an increase in one variable with a decrease in the other (see Figure 5.2).

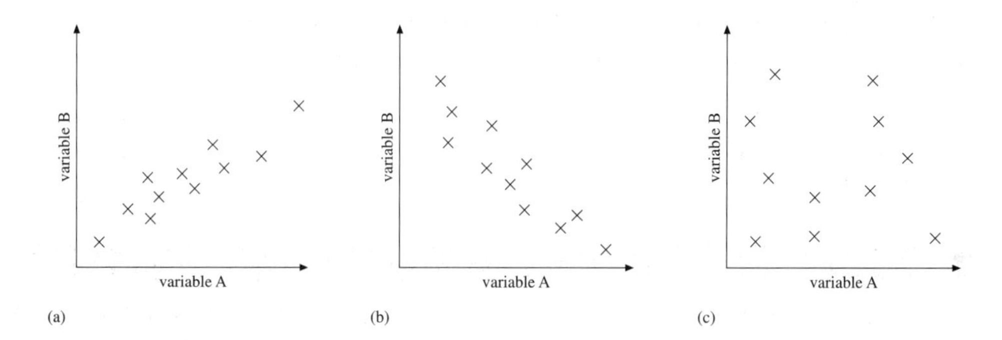

Figure 5.2 *(**a**) Positive correlation (**b**) negative correlation and (**c**) no correlation*

The method of carrying out a Spearman rank correlation test is as follows.

1 To check that it is worth carrying out the test, start by plotting your data on a scattergraph. Plot the independent variable on the x-axis and the dependent variable on the y-axis. The data shown in Figure 5.3 show a relationship. As the velocity of the water in the stream increases, the number of mayflies increases. It shows a positive correlation, so it would be worth carrying out the Spearman rank correlation test. A line of best fit can be calculated using regression analysis. If the data falls into a 'U' shape, then there is no correlation.

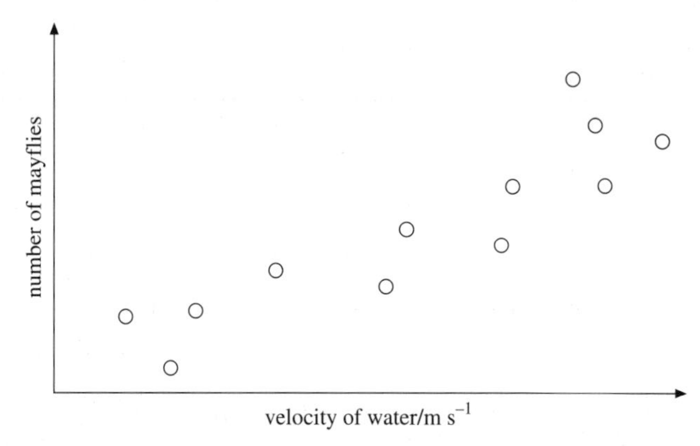

Figure 5.3 *Graph showing the number of mayflies found in water moving at different velocities*

2 The data then needs to be organised in a table. In the first column list the values for the first variable in rank, with the lowest value at the top. In the next column, write the rank for each value, with the lowest value getting the lowest rank. Values which are the same get a mean of the shared ranks. Write in the values for the dependent variable in the third column. Write the ranks for these values in the fourth column.

3 You now have two lists of ranks. Calculate the difference (D) between the ranks, for each pair of measurements. To make sure that you have not made an error, add up all the differences and they should come to zero.

4 Square each of the D values and add them all up.

5 Calculate the correlation coefficient (r_s) using the following formula:

$$r_s = 1 - (6\Sigma D^2/n(n^2-1))$$

6 The resulting value is compared against the critical value. If the value is greater or equal to the critical value, there is a significant correlation between the two variables.

Worked example

A student measured the light level and percentage cover of a species of woodland plant along a transect, which stretched from the middle of a glade into the middle of the wood. He carried out a Spearman rank correlation test on his data (see Table 5.7).

Table 5.7

Quadrat	Light level / arb. units	Rank	Percentage cover	Rank	D	D^2
1	100	9.5	0	1	8.5	72.25
2	100	9.5	5	3	6.5	42.25
3	95	8	4	2	6	36
4	90	7	7	4	3	9
5	75	6	9	5	1	1
6	70	5	41	6	−1	1
7	65	4	45	7	−3	9
8	40	3	66	8	−5	25
9	37	2	71	10	−8	64
10	31	1	67	9	−8	64
Σ					0	323.5

$$r_s = 1 - (6\Sigma D^2/n(n^2-1))$$

$$r_s = 1 - (6 \times 323.5/10(100-1)$$

$$r_s = 1 - 1.96 = -0.96$$

In this case, the critical value for ten samples is 0.648 (the sign is ignored). The calculated value is greater than the critical value, so there is a significant correlation between the two variables.

5.9 Chi-squared

The Chi-squared, χ^2 is a statistic that is used to test an association between two sets of measurements collected in different places (frequency data). The data must be grouped into classes, for example colour, size, sex and the total number of observations must exceed 20. This test is most frequently used in genetics investigations. It has limited use in ecological investigations, for example: are periwinkles associated with a particular species of seaweed?

Chi-squared is a significance test. It tells us whether a relationship is significant or not, and it is based on probability or chance. In order for the relationship to be significant, the chances of the results being due to chance must be low. The acceptable level is 5% significance. This means that the results being due to chance occur only five times in every 100.

The null hypothesis states that there is no relationship and that differences are due to chance. The data are tabulated in the following manner (see Table 5.8).

Table 5.8

Groups	Observed frequency (O)	Expected frequency (E)	$(O-E)^2/E$

$$\chi^2 = \Sigma\,(O-E)^2/E$$

The value of E is calculated by dividing the total number of observations by the number of groups. The degrees of freedom are calculated by subtracting 1 from the number of groups, i.e. $n-1$. A significance level is selected and the χ^2 value is read from a table using the degrees of freedom. The null hypothesis can be rejected if the calculated χ^2 is greater than the critical value.

Worked example

A student had read that the smooth periwinkle, *Littorina obtusata*, was more likely to be found on bladder wrack than on other seaweeds. To test this preference, the student located 100 smooth periwinkles in 25 m^2 sample area of rocky shore. For each periwinkle, the student noted the seaweed on which it was found (Table 5.9).

Null hypothesis: there is no difference in the numbers of smooth periwinkles on the different seaweeds.

Table 5.9

Groups	Observed frequency (O)	Expected frequency (E)	$(O - E)^2/E$
Spiral wrack	2	25	21.16[a]
Egg wrack	30	25	1

table continued ➢

Worked example *continued*

Groups	Observed frequency (O)	Expected frequency (E)	$(O - E)^2/E$
Bladder wrack	61	25	51.84
Serrated wrack	7	25	12.96[a]

[a] ignore the negative sign.

If the periwinkles were randomly distributed, the student would expect to find 25 periwinkles on each of the seaweeds.

$$\chi^2 = \Sigma (O-E)^2/E$$

$$= 86.96$$

Degrees of freedom $(n-1) = 3$

Referring to a reference table of Chi-squared values, the critical value of Chi-squared at the 5% level is 7.82 and at the 1% level is 11.34. The calculated value of 86.96 exceeds this value, so the null hypothesis can be rejected. The periwinkles are not randomly distributed, but showed a preference for certain seaweeds.

5.10 Using computer programs

Carrying out the mathematics of a statistical test can be quite complex. There are a number of software programs that will calculate these values for you. This is perfectly acceptable as you are not going to be tested on the method, but on the significance of the results. If you use a computer make sure you print out a summary of the different stages and indicate whether your data show any significant differences.

CHAPTER SIX

Drawing skills

Several of the awarding bodies include drawing skills in either the course work assessment or as part of a practical exam. Some of the written questions in theory papers set by Edexcel could ask candidates to draw cells from a photograph.

6.1 What makes a good drawing?

The key to a good biological drawing is not to be artistic. It is more important to have good observational skills. You are aiming to produce an accurate record of the features you have observed on a particular specimen. The final drawing should be a neat, simple line drawing of the specimen without shading and other enhancements. Do not be tempted to include features that you cannot observe. You may have seen these features in a text book diagram, but they may not be present or visible in a real specimen, so only draw what you can actually see and not what you think should be there. Often, the best way is to draw the specimen with all your biology books closed so that you are not tempted to check details in a book or to copy from it. It's surprising how many students think they can get away with copying from a book, but they forget that all their teachers have had to produce drawings and they won't be fooled.

6.2 Types of drawings

There are three types of drawing that you could be asked to produce: a drawing of a whole specimen, a low-power plan of a specimen as seen under a microscope and a high-power plan of a few cells of a microscopic specimen.

Low-power plans are usually drawn using the lowest magnification such as ×10 or ×40. The aim is to produce an outline of the areas of different tissues that you can see. You should not add any details of the cells. Don't be worried if your drawing consists of just a few lines. Make sure all the parts are in proportion and add a scale. You can use an eyepiece graticule to help get the proportions right (see Chapter 2).

High-power drawings, as the name suggests, are prepared using the highest magnification of the microscope, for example ×400. Once you have focused on the specimen, choose three or four cells that look typical of the tissue you have to draw. These cells have to be drawn in detail, so look carefully at their shape, the presence of cell walls or membranes and content. As always, the cells have to be in proportion to each other. You may not be able to see all the contents clearly, so don't be tempted to add any non-visible details. For example, you may not be able to see all the chloroplasts in a palisade cell, so just outline one or two that you can see. Also the nucleus may not be visible.

6.3 Labelling your drawing

Some drawing assessments are based on just the actual drawing and not on the labels. If you do have to label your drawing, make sure you draw straight label lines. The labels should be written away from the actual drawing so that they do not obscure any of the detail. Sometimes very brief labels are sufficient. However, you may be asked to annotate your drawing. Annotations are short explanatory comments on points of biological significance. These comments should be concise and relevant. You don't want to cover the paper in annotations.

Checklist of tips for producing a good-quality biological drawing

- ★ Start with a clean piece of paper.
- ★ Choose a hard pencil, such as a 2H, and make sure it is sharpened.
- ★ Have a hand lens available to check details.
- ★ Make sure that your drawing fills up at least half the sheet and is large enough to show all the relevant features.
- ★ Look at the specimen carefully before you start and try to get the proportions right – if you take time here it's easy to fill in more details later.
- ★ Draw a clean, clear outline.
- ★ Add some of the other features that you can see.
- ★ Add straight labelling lines using a ruler.
- ★ Label the features fully.
- ★ Add a scale so that you can give some idea of size.
- ★ Add a title that describes the specimen that you have drawn.

6.4 Sample drawings

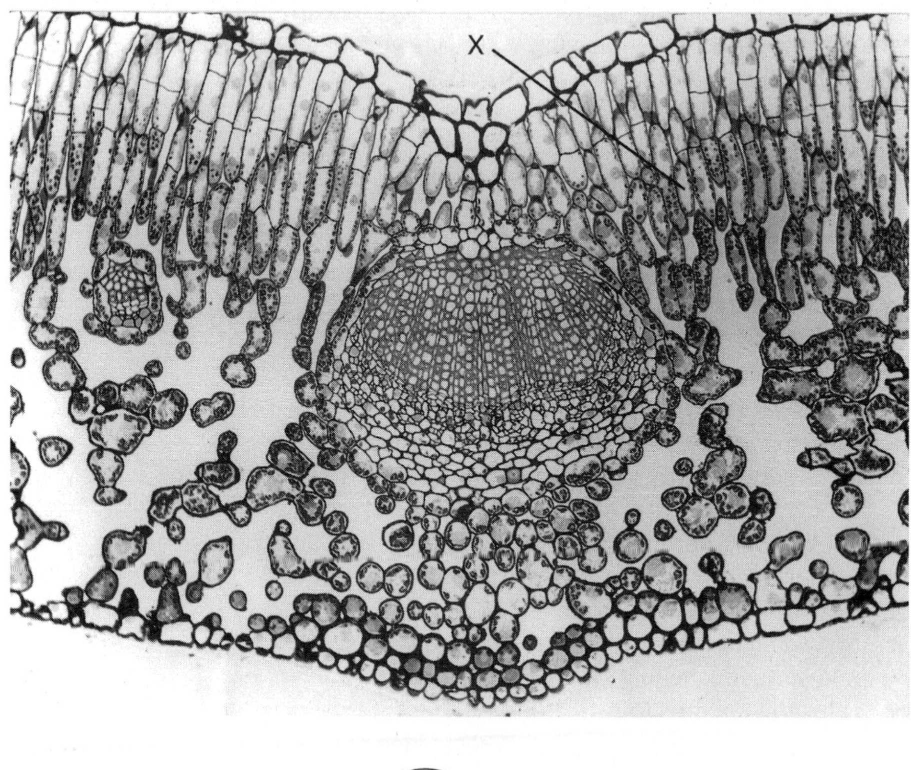

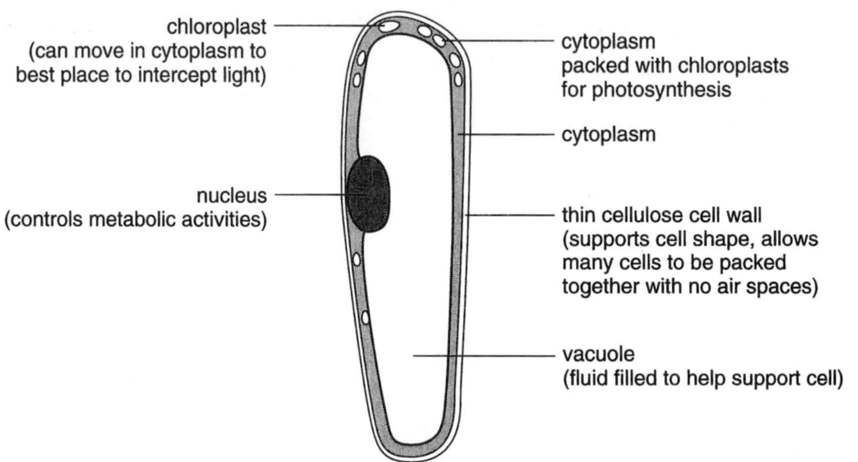

Figure 6.1

The photograph in Figure 6.1 show a section through a leaf. The palisade cell labelled as X has been drawn below. As you can see from the drawing, only those features that are visible in the photograph have been included in the drawing. It is very difficult to see all the chloroplasts, so only a few have been drawn. The cell is drawn to scale with the correct proportions. You may want to draw a few of the palisade cells as a practice to check that you can draw cells with the correct shape and proportions.

Figure 6.2 is a low-power drawing of a transverse section buttercup (*Ranunculus*) root. The drawing fills much of the page and shows the main tissues in outline. The scale is included. The xylem in the middle of the root is shown as a hatched area. This may have been better left unhatched. The lines are smooth and clear. However, this is an accurate representation of what was observed under the microscope.

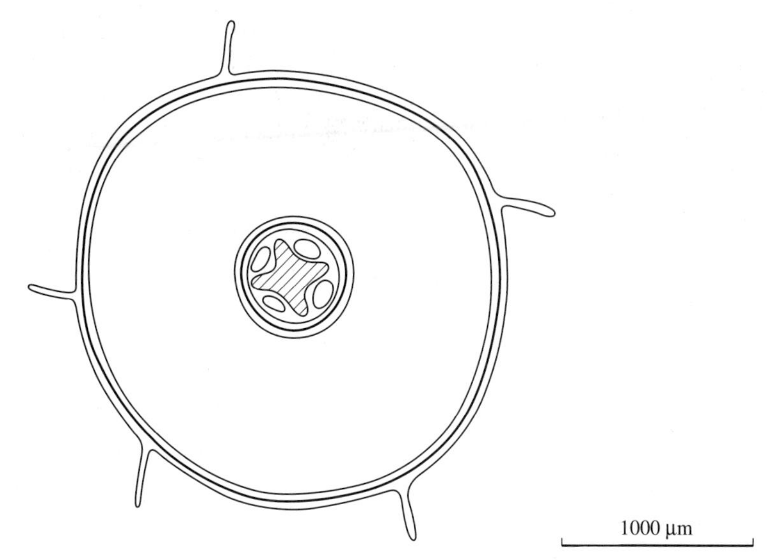

Figure 6.2 *Drawing of a transverse section of a buttercup (*Ranunculus*) root*

Figure 6.3 is a high-power drawing of four cells from phloem tissue. The drawing is large and only a few cells have been drawn. The shape of the cells is accurate and there is a scale. The student has stippled those areas that look darker under the microscope. The sieve plate is difficult to draw and so only a few of the sieve pores have been added. Overall this is a good-quality drawing.

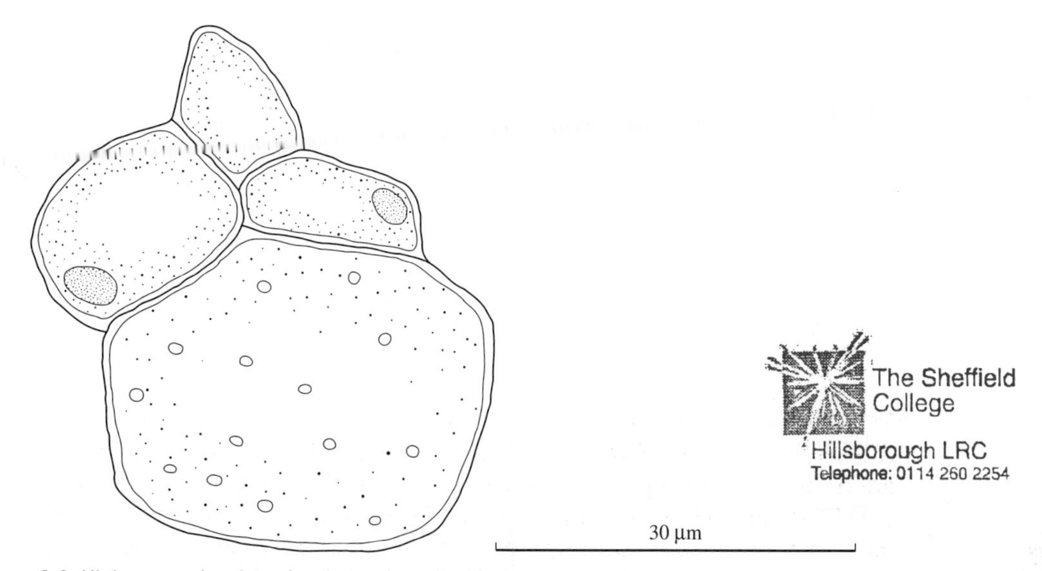

Figure 6.3 *High power drawing of a sieve tube cell with three companion cells*

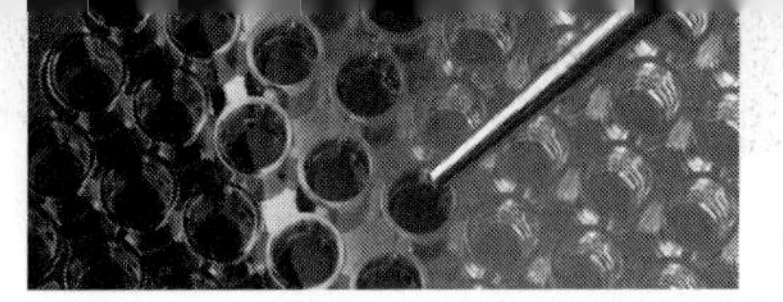

CHAPTER SEVEN

Individual investigations

Students carrying out A2 coursework need to show some progression from AS, so although the skills are similar, they are assessed at a higher level. Most of the awarding bodies require A2 students to carry out an individual investigation, in which various elements are assessed. The skills include planning, implementation, analysing, evaluating and report writing. The style of the report, including the biological introduction, may also be assessed.

The choice of your investigation will be critical in determining your final marks. The investigation has to be suitable for carrying out measurements and for a statistical test to be applied to the data. Many of the boards require the A2 investigation to include information drawn from different parts of the specification. In other words, your investigation needs to be synoptic in some way. For example, in your discussion of the results you could include information from the AS and A2 specifications.

7.1 Planning skills

These are similar to those outlined in Chapter 1, but they are assessed at slightly higher levels; for example you would be expected to be able to fully justify your choice of measurements and ensure that they are suitable for statistical analysis. You would need to indicate how you would control all of the variables. The level of research into the problem may be greater, and you may need to draw upon information and skills taken from different parts of the AS and A2 specifications.

7.2 Implementing

You will be expected to be competent in handling of the apparatus. At this level, you would be expected to show a wide range of manipulative skills, and any measurements would have to be of a high degree of precision with attention to detail.

7.3 Introduction

Some awarding bodies require students to write an introduction to their investigation. The introduction identifies and defines the problem and includes background biological knowledge. The aim is to produce a concise and accurate introduction that informs the reader. It is not necessary to write a lengthy essay on the subject. The introduction may include references to prior knowledge and published references that have be found in books and on the Internet.

7.4 Method

A2 students may be expected to write a report of the method they used in their investigation. This skill is not normally assessed at AS. The report will have to be written in continuous prose, rather than as a simple list of points. All of the experimental details have to be included, so that it would be possible for another person to repeat the investigation and obtain similar results. As mentioned earlier, the report must be written in the past tense, but there is no requirement to write in the third person. You should write in the style that you find easiest. The report should also refer to any precautions that were taken.

7.5 Analysing

A2 students will be expected to be competent in producing tables and choosing appropriate graphs and charts to display the trends. All the trends and patterns observed in the data would be expected to be noted and commented upon. In addition, the data should be tested statistically using a null hypothesis. Comments should be made on any anomalies in the results.

7.6 Discussion and evaluation

The discussion should be clearly presented and the data used as evidence to support the hypothesis. It should be of greater depth than those produced at AS level. It should show good biological knowledge and include logical explanations of the results using the correct terms. Information should be drawn from different parts of the specification in order to achieve the synoptic element of the assessment. There will be some discussion of the limitations and how these limitations may have affected the results. There should be some proposals for further study.

7.7 A2 investigations

When planning an investigation, try to choose a subject that interests you. You may get ideas for the investigation by reading magazines or from watching television programmes. You will probably carry out your investigation towards the end of your A2 course, so make notes of any possible subjects to investigate that you come across while working through the AS specification. Some of the research into the background of your chosen subject could be done earlier in the course too.

Some of the features of a good investigation include:

- novel idea rather than the modification of a basic textbook experiment
- displays a high level of manipulative skills
- generates good-quality quantitative data
- generates data that can be analysed statistically
- allows the discussion to be synoptic.

It is impossible to give a comprehensive list of the types of investigation that could be carried out. However, the paragraphs below give a few 'starter' ideas. Some of the ideas are simply observations that you could develop. Others are starting points with some guidance on the types of tests that you would need to carry out. Remember that you will have to carry out a statistical test. This will be one the major factors in your experimental design.

Investigation topics

1. Cell biology and enzymes

As part of the AS specifications you will have carried out experiments using common enzymes such as catalase and amylase. An A2 investigation has to go further than simply looking at the effect of temperature, pH, substrate and enzyme concentration on a rate of reaction. An enzyme investigation could be raised to the A2 level by including an A2 topic such as metabolism, for example, looking at the effect of respiratory inhibitors such as heavy metals on enzyme activity. You may want to avoid the enzymes you used in class, for example amylase and catalase, and focus on some of the commercial enzymes. Enzymes are used in biological washing powders, in the production of fruit juices, in removing lactose from milk, producing sweeteners for the food industry and much more. Lactase is an enzyme that is easily immobilised. You could study the effect of different sizes of beads, or the rate of flow through the beads on the conversion of lactose to glucose and galactose. Processes such as making cheese, beer and yoghurt all involve enzymes.

Pectin is a substance that helps to hold plant cells together. As fruits ripen, the pectin breaks down and they become softer. The change is due to the production of pectinases. Commercial pectinases can be purchased for use in jam making. This substance could be used in an investigation into the effect of pectinase in fruit ripening or in juice extraction.

Pineapples contain a protease enzyme called bromelain, which increases in concentration as the fruit ripens. An investigation could be based around this fact and pineapples of different ages could be tested. Alternatively, bromelain could be investigated under different conditions using gelatin, which contains protein.

The enzyme, ascorbic acid oxidase, is released from fruits when their tissues are damaged. This enzyme is activated on exposure to air and it causes the tissue to lose ascorbic acid (vitamin C). The enzyme is inactivated by boiling the juice as soon as it is extracted from the fruit in order to retain the vitamin C content. This could form the basis of an investigation into how effective this process can be and whether all fruits react in the same way. Apples contain another enzyme, diphenol oxidase, which causes the flesh of the apple to discolourise on exposure to air. The presence of ascorbic acid (vitamin C) in the apple will inhibit this enzyme.

The ascorbic acid content (vitamin C) and sugar content of foods such as fruits are affected by the conditions under which the foods are kept. The concentrations of reducing sugar can be determined semi-quantitatively using Benedict's reagent and a range of standard glucose solutions. Ascorbic acid content is determined using the indicator, DCPIP (phenol-indo-2,6-dichlorophenol). First, the volume of a standard solution of ascorbic acid required to decolorise a fixed volume of DCPIP is determined. Then this is repeated with the ascorbic acid replaced by an extract from the fruit.

Yeast can be used as the basis of many different investigations. You will need to set up a method by which you can estimate the number of yeast cells, for example by using a colorimeter or haemocytometer. There are many different strains of yeast, each suited to a different role. For example, there are baking and brewing strains. You could grow different types of yeast on a number of different carbon and nitrogen sources, such as fructose, glucose, sucrose, etc. You could also investigate rates of fermentation by different types of yeast or study the effect of alcohol (ethanol) on the growth rate of yeast. Bacteria are involved in yoghurt formation. You could investigate the optimal conditions for its production using changes in viscosity or pH as a means of monitoring the process.

If you are studying a respiring suspension of yeast, you may need to use a redox indicator such as tetrazolium chloride (TTC) or methylene blue. These two chemicals are artificial hydrogen acceptors and they change colour as they are reduced by hydrogen ions. TTC changes to pink as it is reduced, while methylene blue becomes decolorised.

2. Human physiology

Exercise can affect heart rate, vital capacity, body mass and fat. Investigations could look at the effect of regular exercise on the resting heart rate, recovery periods and general fitness of an individual or on their body mass index. It is important to stardardise the exercise activity, for example using an exercise bike or running a fixed distance. The heart rate is also affected by stimulants such as nicotine and caffeine. Heart rate can either be measured direct by using a stethoscope or by taking a pulse. The accuracy of pulse taking can be improved by using digital pulse monitors.

It is also possible to monitor the effect of a range of different chemicals, such as alcohol, aspirin, caffeine on the heart rate of the water flea, *Daphnia*.

The vital capacity of the lungs varies between individuals. Often athletes and singers have a large vital capacity. Vital capacity can also be improved with aerobic exercise. Investigations into vital capacity could be linked to heart rate investigations. Vital capacity can be measured by asking the subject to breathe out into a special plastic bag fitted with a mouthpiece and a scale.

Studies may be made into the change in reaction times to various audio and visual stimuli. Reaction time is measured using the 'ruler-drop' test. A metre rule is held at the 50 cm mark between the thumb and forefinger by the experimenter and dropped without warning. The subject of the investigation has to grip the ruler as it passes through his/her hand. The distance that the ruler drops is converted to time using the following formula:

$$t = \sqrt{2s/g}$$

where t = time (seconds), s = distance dropped (metre), g = 9.81 (acceleration due to gravity).

Also it is possible to use computer programs to carry out reaction time tests. Investigations can determine whether reaction time is affected by sex, age, time of day, fatigue, etc.

The quantity of urea in urine can be determined using the enzyme urease. The urea is broken down into ammonium carbonate, which can be estimated by titration against hydrochloric acid. Methyl orange is used as the indicator. Investigations can be carried out into the link between diet and the quantity of urea in urine.

Remember that if you are using human subjects in your study, you need to do a thorough risk assessment (see the detailed information provided on pages 9–11).

3. Plant physiology

Chloroplasts can be isolated by taking plant tissue, blending it in a buffer solution, filtering and then centrifuging the filtrate to produce a pellet of chloroplasts. The chloroplasts can be resuspended and investigated. For example, it is possible to investigate the photolytic breakdown of water to produce oxygen (often called the Hill Reaction). The chloroplasts are mixed with redox indicator such as DCPIP and placed in bright light. A colorimeter is used to plot the change in colour from blue to colourless. The control is produced by boiling the chloroplasts.

This basic technique can then be applied to a number of situations. For example, the effect of different kinds of weedkillers on isolated chloroplasts, the effect of different light intensities on chloroplasts taken from different types of plant, or from plants of the same species growing in different situations, or the effect on shade- and sun-loving plants.

The textbook method for measuring the rate of photosynthesis is to determine the rate of production of oxygen from an aquatic plant. The apparatus is called a photosynthometer. The bubbles coming off the cut end of a piece of pondweed are collected and funnelled into a capillary tube. This way, the length of the bubble can be measured. Changing variables, such as light intensity, wavelength and carbon dioxide concentration, is not really suitable for the individual investigation. But you could study the effect of different wavelengths of light on photosynthesis in different types of seaweeds and link this to their position on the shore. Also you could add a weedkiller to the water around the pondweed. You could look at the effect of acidifying the water and link this to acid rain damage.

Another useful plant is the duckweed, *Lemna minor*. Duckweed is a tiny flowering plant. It consists of small floating fronds or leaves with one or two roots. It is found growing over the surfaces of ponds in summer. It grows very fast and can be used to study the effects of various minerals and pollutants on plant life. It can be grown in Petri dishes and measurements can be made of frond reproduction, increases in mass or the total quantity of chlorophyll. It can be grown in a standard nutrient solution that contains all of the minerals required for healthy growth. The investigation can be varied to look at the effect of nutrient deficiencies, the presence of heavy metals such as lead, zinc, copper and chromium, or chemicals such as nitrate and phosphate fertilisers, detergents or even bleach. Chlorophyll can be extracted from the fronds using 80% ethanol or *N,N*-dimethyl formamide (DMF).

A potometer is used to measure water uptake by a plant shoot. The rate of water uptake can be linked to the rate of transpiration. You may have used this apparatus in a class experiment to look at the effect of wind speed, humidity, etc on transpiration. It is possible to use this experiment as the basis of an investigation. You could study transpiration in a number of different species of plant, for example those growing in woodland and those growing in a more open situation, and then link the results obtained with the potometer to microscopic studies looking at epidermal strips from the chosen species. Cut flowers are often provided with a sachet of powder that contains sugar and aspirin to extend the life of the flowers. You could carry out an investigation to determine the effect of aspirin and sugar on the rate of transpiration.

Seed germination is a good topic to study. You could study the effect of light intensity, wavelengths of light and temperature on the percentage germination of a range of different seed types. Some seeds require exposure to high or low temperatures before they will germinate, while others need to be soaked in order to wash out germination inhibitors. Seeds are often stored for long periods of time and it is important to know if the seeds are still viable. There are various tests of seed viability. Many seeds secrete sucrose while they are viable and this can be detected with Benedict's reagent. Another test is to cut the seeds in half and soak them in a solution of tetrazolium salt. Viable seeds stain red because they contain catalase. The level of catalase within the seed decreases as the seeds age. You could use the catalase–hydrogen peroxide reaction to test for viability. Any of these tests could be used in an investigation that studies the effects of different storage methods on the viability of different types of seed.

Plant growth opens up all sorts of possibilities. Using easily obtained plants such as cress, wheat, peas and beans you could set up investigations to look at the effect of different types of fertiliser on plant growth. Many schools grow plants called fast-cycling brassicas. These are plants that have been developed for use in schools. As the name suggests, the growth rate of these plants is very fast and many complete their life cycle within 6 weeks. They are grown under bright lights. These plants allow students to complete studies on plant growth within a half term. If your school does have these plants you could check out the SAPS website for ideas.

In autumn, deciduous tree leaves undergo a change in colour before they fall. The changes in colour are linked to the progressive breakdown in chlorophyll within the leaves. Trees of the maple family undergo particularly colourful changes. This is supposedly linked to the higher levels of sugar in the tree. The colour changes in various tree species could be investigated, using chromatography. This would enable the different pigments within the leaf to be separated and identified.

A number of investigations can be carried out into plant growth regulators such as auxin and giberellin. For example, does hormone rooting powder lead to more rapid root formation in a pot plant such as the geranium? There are several different types of hormone rooting powder. Is one formulation any better than the others? Cuttings can be placed in water or grown in media such as vermiculate, peat or sand. Does the rooting medium affect the formation of roots in stem cuttings?

4. Ecology and environmental topics

There are numerous possibilities in the field of ecology. However, you have to be careful that your investigation does not get too big. It is better to plan an investigation that focuses on a particular species. One way to source ideas is to walk around a local woodland, park or grassland. You may notice that a certain type of plant is only found in one part of the area. Abiotic factors such as soil type, pH or moisture and light intensity could all affect its distribution. Sometimes, the effect of trampling by walkers can affect plant distribution. Ecological investigations could include a number of skills and techniques such as sampling, line transects, use of quadrats, use of keys to identify plant species, laboratory skills in analysing soil water, oxygen, mineral content, etc, plotting species profiles and species distribution maps.

Sampling will be an important aspect of your plan. Will you carry out random sampling in the areas of interest, or will you sample a transect across the area? A transect would allow you to monitor abiotic factors at each sample point to determine if there is link between distribution of the plant and an abiotic factor. But beware – the types of statistical test that could be carried out are limited. You will probably have to carry out a correlation. Sometimes, there are associations between certain species of plant or between plants and certain animals, for example thyme is often observed growing on anthills. This type of relationship could be tested using a correlation or by using a two-sided Chi-squared test.

There are many possibilities along a rocky shoreline. You could look at the height and diameter of mollusc shells and relate this to the exposed or sheltered nature of the shore. You could take samples of different types of seaweed and carry out a photosynthesis-based investigation. Some seaweeds are more tolerant of desiccation than others. You could dry out seaweed fronds for specific periods of time and, on rehydration, test the rate of photosynthesis as a measure of their recovery.

The types of invertebrate animals found in a pond and their abundance are affected by many factors, for example, the age of the pond, the pH, oxygen content, nitrate levels etc. You may be able to locate ponds of different ages and see if there are any differences in the animal diversity, or find ponds that have been polluted with farm run-off or fertilisers and compare them with ponds that appear to be unpolluted. Ponds surrounded by trees are shaded in summer and in autumn lots of leaves fall in the water. This can affect the pond life. You could look at animal diversity and abundance in shady ponds and in ponds in an open location. You could also compare garden ponds with those in parks and on farmland.

You could look at the effect of various factors such as pH, nitrate level, etc, on the growth of pond organisms, for example *Daphnia*. Ponds and rivers that suffer from high levels of nitrate often suffer from algal blooms. You could devise a method for looking at the growth of filamentous algae in water that varies in nitrate and phosphate levels.

Acid rain has been reported as having a negative effect on plant growth. You could grow plants using acidified water and observe the effect of the acidity on their growth.

Some lichens are particularly sensitive to the levels of sulphur dioxide in the atmosphere and can be used as indicators to the level of acid rain pollution in an area. You could find out which types of lichen are sensitive and which are not, and survey the lichens growing on trees and walls in a city or near an industrial area to a more rural location. You would need to devise a method for sampling the lichens. You may even want to calculate an index of diversity.

The air pollution produced by car exhausts can affect plant growth. This could be investigated to see how the pollution affects the plants growing on roadside verges. This could be linked to photosynthesis and enzymes. It has also been noted that plants growing near roads support larger populations of aphids than the same plants growing further from the road.

Leaves of plants often accumulate unpleasant tasting chemicals, called tannins, to discourage insects from eating them. In general, older leaves have more tannins than young leaves. The presence of tannin can be tested by boiling sample material in water and then adding a solution of iron chloride to the boiled sample. The intensity of the resulting black colour is an indicator of the concentration of tannin in the leaf. The results of these types of tests could be combined with observations in the field.

Many tree seeds do not germinate when they fall beneath the adult tree. This could be due to chemicals that are present in the ground. It is also observed that the ground beneath certain trees, such as the walnut, lacks vegetation. Again, a chemical washed from the leaves or from the tree roots could be responsible for the lack of plant growth.

Example: A2 investigation

The investigation below is one that was prepared for the Edexcel specification. Although some of the sections may not be applicable to all boards, it should give all A2 students some idea of the levels they have to achieve. A commentary is given at the end.

Investigation of the common limpet (*Patella vulgaris*)

Abstract

The investigation studied the common limpet, *Patella vulgaris*, living in the lower, middle and upper shore of the exposed and sheltered sides of Watwick Bay in Wales. Measurements were made of the maximum height and diameter of 30 limpets found systematically on the lower, middle and upper shores of each beach. The ratio of height to diameter was calculated. The standard deviations of the data sets were obtained and a *t*-test was carried out on some of the sets. There was an overall trend for the height of the shell to increase from the lower shore to the upper shore on both beaches. The only significant differences were found for the height measurements and ratios of limpets living on the upper shores compared with the middle shore. There was also a difference between the upper shore of the exposed and sheltered beaches. It was concluded that limpets living on the upper shore were exposed for greater periods and they had to clamp themselves firmly to the rocks to avoid desiccation. This meant that their foot was contracted and smaller, and a taller shell was secreted.

Aim

To investigate the variation in the vertical height and greatest diameter of shells of the common limpets living in the lower, middle and upper shore of the exposed and sheltered sides of Watwick Bay.

Null hypothesis

There will be no difference in the mean height, diameter and ratio of height to diameter of the shells of a common limpet living in different parts of the shore or on an exposed and a sheltered beach.

Plan

I will first obtain the tide times. From my initial studies I know that the high tide on the day I wish to carry out the investigation is 6.7 m above datum.

I will first have to mark out the upper, middle and lower shore on the exposed beach. I will take the upper shore as being 5 – 7 m, the middle shore is 5 – 3 m and the lower shore is 3 m to the sea. I will determine these measurements using a metre rule. I will place the metre rule vertically at the lowest sampling point and place a 30 cm ruler horizontally across the top of the metre rule. I will use a spirit level to make sure that the short ruler is horizontal. I will look along the ruler to pick out a point 1 m above my sample point. I will repeat this up the beach. I will mark the position of the boundaries of the upper, middle and lower shore by placing rulers on the rocks as markers.

I will sample 30 limpets in each of the shores of the beach. I will sample the limpets systematically. I will lay a tape measure up the beach from the lowest point to the highest point. Every 10 cm, I will measure the limpet that is nearest to the left side of the tape. The height and maximum diameter of the limpet will be measured using a vernier caliper. This will enable me to get accurate readings. The calipers will be more precise than using a ruler. Once back in the laboratory, I will calculate the ratio of height : maximum diameter and determine the frequency of the different classes. I will plot graphs of frequency against height, diameter, and ratio of height to diameter, for each zone on the exposed and sheltered shores.

I will have to carry out a pilot study to determine whether my data is of normal distribution, or skewed, as this will affect my choice of statistical test. If the data has a normal distribution, I will calculate the mean and standard deviation. I will also carry out a Student's *t*-test. If the data is skewed, I will carry out a Mann–Whitney *U* test.

I will have to control as many variables as possible. I am investigating two independent variables – firstly, I am looking at the difference in the position on the shore and secondly, I am looking at the exposure of the beach. The dependent variables are the height and diameter of the shells. All other factors have to be kept as constant as possible. The two shores that I have chosen are very similar. They have the same size and types of rocks. They differ in aspect and direction of prevailing winds and waves. I am measuring the height above sea level to make sure that the samples are all taken in the correct shores. I will be sampling at the same time of day and at the same point in the tide cycle. The light levels should be approximately the same, as the samples are taken at the same time of year. I cannot control aspects such as cloud cover, wind and rain, which may differ on the sampling days. However, I don't think that these environmental variables will affect my results, as the shells take some time to grow and the size will not be affected by differences in weather on two consecutive days in summer.

Since I will be working up from the lower shore, I will not have time to carry out the sampling on both beaches on the same day safely. So, I will carry out the sampling on the sheltered beach on the following day. The tide times will alter by just under 1 hour. This means that low tide will be at a different time of day. Hopefully, the difference of less than 1 hour will not affect the results.

Equipment

30 m tape
1 × metre rule
1 × 30 cm ruler
small spirit level
vernier caliper to measure limpets
pad, board and pencil for recording measurements.
a number of brightly coloured sticks to act as markers
watch (to monitor tide times)

Pilot study

I measured the height and diameter of 20 limpets shells at random taken from approximately the middle shore using a vernier caliper. I found that it was quite difficult to measure the height of the shell using a vernier caliper so I will use two short rulers held together with an elastic band to form a small measuring stick. One of these rulers will stand vertically beside the limpet and the second will be held at right angles to the first so that it lies across the top of the shell. I will be able to read off the level where the second ruler crosses the first. Unfortunately, this arrangement will not produce as accurate a reading as the vernier caliper but it is the best method, given the circumstances.

Back in the laboratory, I calculated the ratio of height to maximum diameter. I then produced a quick tally chart (Table 7.1) to see if the values had a normal distribution. The data did have a normal distribution so I decided that the *t*-test would be an appropriate statistical test.

Table 7.1 Table showing the height : diameter and frequency results of the pilot test

Height : diameter	Frequency
0.1	0
0.2	2
0.3	5
0.4	7
0.5	4
0.6	0
0.7	1
0.8	1
0.9	0
1.0+	0

Risk assessment

No chemicals or other harmful substances will be used in this investigation.

I will carry out my investigation while the tide is going out. I will check the time of low tide with the warden. This means that I will not be working on the lower shore with the tide coming in, which is dangerous as I could be cut off by the tide. I will be working on a rocky shore where there are rock pools and sharp rocks covered with seaweed. I will take care not to slip on the rocks and wear Wellington boots to protect my feet.

I will be handling living organisms. I will take care not to damage them while measuring them and I will try not to crush too many animals as I clamber over the rocks.

Introduction

Limpets are molluscs. There are two types of limpet found around Dale. They are *Patella vulgaris* (the common limpet) and *Patina pellucida* (the blue-rayed limpet). The one I will be investigating is the common limpet, *Patella vulgaris*. Limpets belong to the class Gastropoda as they have a single shell and a foot. Unlike most gastropods, the limpet has a simple cone-shaped shell with no spiral. The common limpet has a tall shell up to 6 cm across and ribbed, with an internal surface of white or yellow. Limpets grow in size and their shells grow to fit a particular spot on the rock. This is so that, when exposed at low tide, they are able to clamp their shells down close to the rock avoiding unnecessary water loss. They are found at most zones up the shore, especially near the barnacle zone. They feed on algae, diatoms, and on lichens found on the rocks. They have a tongue called a radula. They are active when below water and move about searching for food in the vicinity of their special place on the rocks.

Limpets are adapted to their zone on the beach. Limpets of the lower shore appear to have a larger and flatter shell than those at the top of the shore. On the lower shore limpets don't have to resist so much desiccation; therefore they do not have to contract their foot that much and so they take up more space and a larger shell is secreted. On higher shores, their feet are contracted to help them hang on to the rock when buffeted by the waves. Their shell is much taller with a smaller diameter.

I will be examining whether the position of the limpet on the shore affects the size and diameter of its shell. Also, I will see if limpets living on a more exposed beach have a different-sized shell to limpets of the same species living on a sheltered beach.

Method

The site of my investigation was Watwick Bay on the Pembrokeshire coast. The bay has rocky beaches. The more northerly beach is exposed to prevailing winds from the south west and is also open to waves that have very long fetches. The southerly beach is protected from the winds and waves and so is more sheltered (see map). It was decided that the shore would be divided up into three zones (upper, middle and lower shore) on each side of the bay.

The low tide for 12th May was 1.0 m high at 13.54 hours. The investigation was started at 12.30 pm when the tide was 2.0 m above chart datum (CD) (CD = range × factor + height of low tide). High tide was 6.7 m above CD. The lower shore was taken as 3 m down to sea level, the middle shore was 3 – 5 m and the upper shore was 5 – 7 m. These measurements were determined using two rules, a short 30 cm ruler and a longer metre rule. The metre rule was placed vertically at the water's edge and the smaller ruler was held horizontally on the top of the metre rule. A spirit level was used to make sure the ruler was horizontal. By looking along the top of the short ruler, it was possible to pick out a point 1 m higher on the shore. This point was marked and the process was repeated. Then the shore was physically divided up by placing marker sticks on the rocks that were on the boundaries of the various zones of the shore.

Then a tape measure was placed up the beach, from the sea's edge to the high tide mark. Within each zone, 30 limpets were measured for height using the two short rulers joined together and for maximum diameter using a vernier caliper. The limpets were selected systemmatically, using the tape. At 10 cm points along the tape, the limpet nearest to the left side of the tape was selected for measurement. This was repeated until 30 had been sampled within each zone. It was impossible to measure limpets under water in rock pools, so the next nearest limpet out of the water was measured. Observations were also made on the appearance of the rocks and on the quantity of seaweed covering the rocks.

The following day, when the tide was at the same point, the procedures were repeated and measurements were made of the limpets on the sheltered beach. Figures 7.1, 7.2 and 7.3 compare limpet heights on exposed and sheltered beach from lower shore, middle shore and upper shore, respectively.

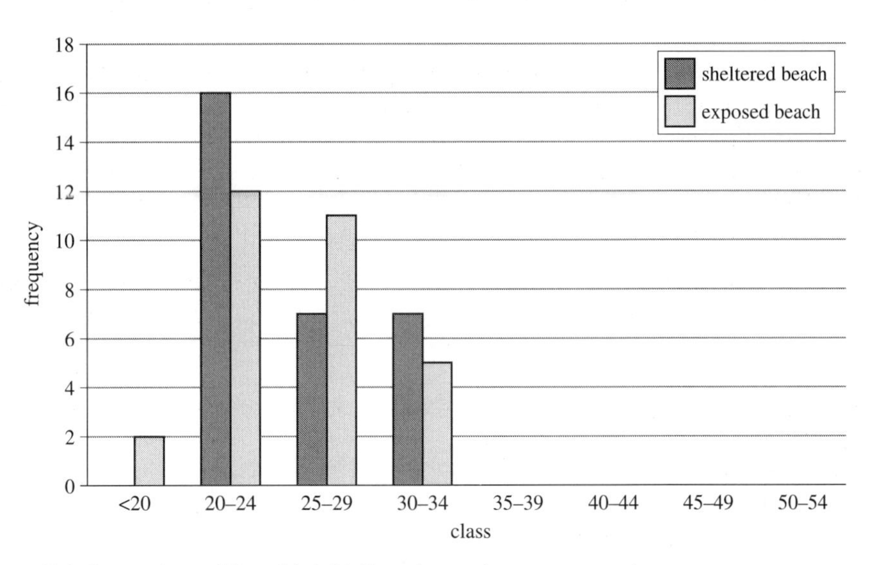

Figure 7.1 *Comparison of limpet heights from lower shore on exposed and sheltered beach*

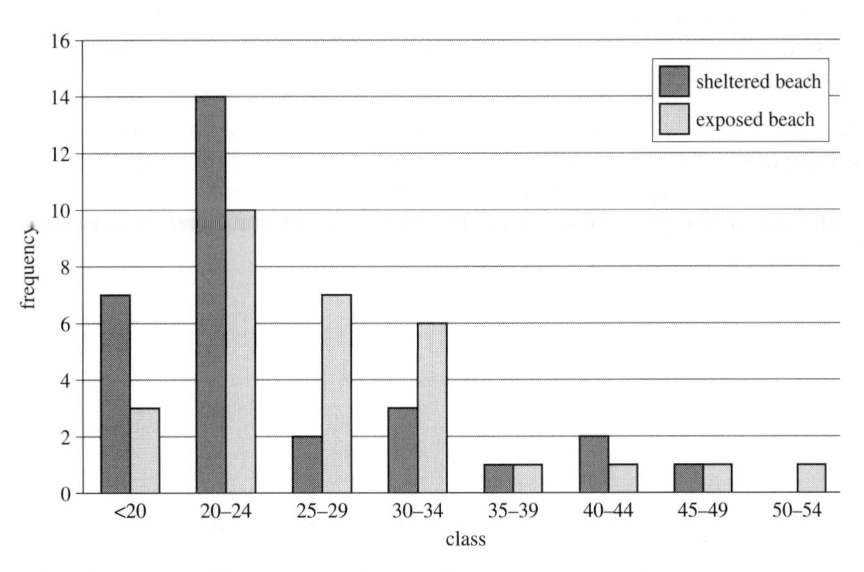

Figure 7.2 *Comparison of limpet heights from middle shore on exposed and sheltered beach*

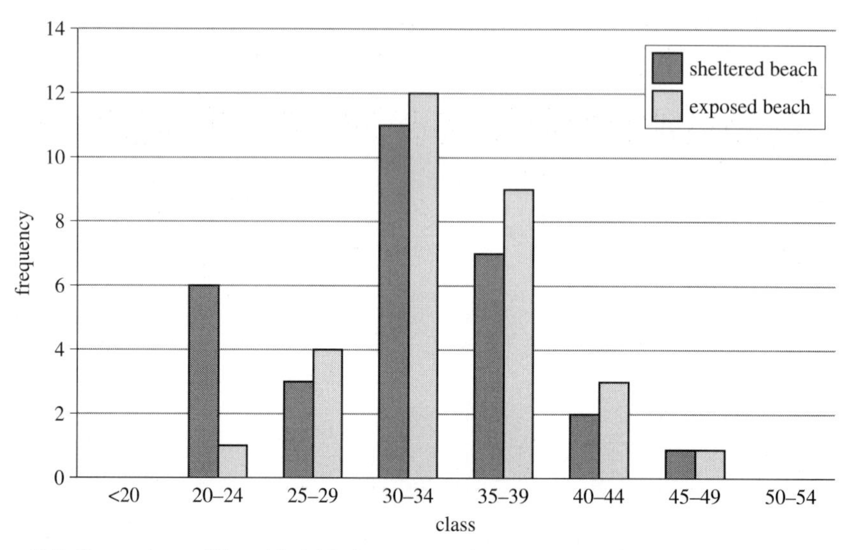

Figure 7.3 *Comparison of limpet heights from upper shore on exposed and sheltered beach*

Table 7.2 *Shows the frequency for different classes of height of shell of Patella vulgaris on lower, middle and upper zone of the exposed and sheltered beaches*

	Height of limpet shell/mm					
	Sheltered beach			Exposed beach		
Class/mm	Lower	Middle	Upper	Lower	Middle	Upper
<20		7		2	3	
20–24	16	14	6	12	10	1
25–29	7	2	3	11	7	4
30–34	7	3	11	5	6	12
35–39		1	7		1	9
40–44		2	2		1	3
45–49		1	1		1	1
50–54					1	

Table 7.3 *Shows the frequency for different classes of maximum diameter of the shell of Patella vulgaris on lower, middle and upper zone of the exposed and sheltered beaches*

	Maximum diameter of limpet shell/mm					
	Sheltered beach			Exposed beach		
Class/mm	Lower	Middle	Upper	Lower	Middle	Upper
35–39			2			
40–44	1	3	6	1	2	2
45–49	9	7	12	6	5	11
50–54	11	4	3	10	7	4
55–59	9	13	6	12	14	12
60–64		3	1	1	2	1

Table 7.4 *Shows the frequency of different classes for the ratio of height : maximum diameter of the shell of* Patella vulgaris *on lower, middle and upper shores of the exposed and sheltered beaches*

	Ratio of height : maximum diameter					
	Sheltered beach			Exposed beach		
Class	Lower	Middle	Upper	Lower	Middle	Upper
<025					1	
0.25–0.29				1		
0.30–0.34		4		1	1	
0.35–0.39	2			1	1	
0.40–0.44	5	6		6	8	
0.45–0.49	7	9	1	9	3	1
0.50–0.54	9	6	5	6	8	3
0.55–0.59	4	5	6	5	2	1
0.60–0.64	3	1	3	1	1	7
0.65–0.69			3		2	5
0.70–0.74			6			4
0.75–0.79			1		1	3
0.80–0.84			1		1	3
0.85–0.89			2			2
0.90–0.94			1			
0.95–0.99			1			
1.00>					1	1

Analysis of results

The results were quite varied. Looking at the tables of frequency (Tables 7.2–7.5), it appears that there is an overall trend for the height of the shell to increase from the lower shore to the upper shore on both beaches. No trend was visible with the maximum diameter data. Also, there was a trend for the ratio to get larger, from the lower to the upper shores. There was also a wider range of values for the upper shore data on both beaches. For example, on the lower shore of the sheltered beach the range of the ratios was 0.38–0.64, but on the upper shore of the same beach the range was 0.47–0.95.

Table 7.5 *t-test: Lower shore comparison of ratios*

	Sample	Sheltered	Exposed
	1	0.38	0.41
	2	0.6	0.47
	3	0.39	0.51
	4	0.53	0.29
	5	0.58	0.41
	6	0.10	0.00
	7	0.47	0.57
	8	0.45	0.49
	9	0.43	0.59
	10	0.47	0.49
	11	0.55	0.53
	12	0.51	0.43
	13	0.56	0.49
	14	0.64	0.34
	15	0.49	0.53
	16	0.53	0.53
	17	0.51	0.49
	18	0.48	0.6
	19	0.41	0.55
	20	0.57	0.58
	21	0.41	0.43
	22	0.49	0.44
	23	0.47	0.43
	24	0.51	0.56

table continued ➤

	Sample	Sheltered	Exposed
	25	0.43	0.47
	26	0.49	0.53
	27	0.53	0.48
	28	0.53	0.48
	29	0.61	0.49
	30	0.51	0.53
Sum of values		14.96	14.52
Mean		0.50	0.48
SD		0.066	0.07
Variance		0.004	0.005
Difference between means		0.015	
Number of samples		30.00	30.00
Mean variance		0.0001	0.0002
Joint variance		0.0003	
t		0.82	

Critical value of t at 5% = 2.042, therefore calculated less than critical, so accept null hypothesis.

Table 7.6 *t-test: Middle shore comparison of ratios*

	Sample	Sheltered	Exposed
	1	0.71	0.44
	2	0.44	0.41
	3	0.46	0.47
	4	0.51	0.68
	5	0.68	0.44
	6	0.80	0.53
	7	0.91	0.35
	8	0.31	0.57
	9	0.37	0.23
	10	0.43	0.69
	11	0.34	0.52
	12	0.42	0.53
	13	0.58	1.02

table continued ➢

	Sample	Sheltered	Exposed
	14	0.47	0.50
	15	0.45	0.53
	16	0.51	0.43
	17	0.40	0.56
	18	0.32	0.41
	19	0.38	0.73
	20	0.43	0.82
	21	0.49	0.40
	22	0.52	0.41
	23	0.40	0.53
	24	0.35	0.47
	25	0.39	0.32
	26	0.39	0.52
	27	0.43	0.61
	28	0.33	0.41
	29	0.61	0.49
	30	0.51	0.53
Sum of values		14.339	15.5476
Mean		0.48	0.52
SD		0.143	0.15
Variance		0.020	0.024
Difference between means		0.040	
Number of samples		30.00	30.00
Mean variance		0.0007	0.0008
Joint variance		0.0015	
t		1.05	

Critical value of t at 5% = 2.042, therefore calculated less than critical, so accept null hypothesis.

Table 7.7 t-test: *Upper shore comparison of ratios*

	Sample	Sheltered	Exposed
	1	0.58	0.66
	2	0.73	0.43
	3	0.53	0.58
	4	0.71	0.67
	5	0.73	0.63
	6	0.63	0.49
	7	0.71	0.53
	8	0.47	0.67
	9	0.82	0.33
	10	0.53	0.64
	11	0.95	0.23
	12	0.51	0.51
	13	0.67	0.71
	14	0.67	0.61
	15	0.51	0.32
	16	0.84	0.79
	17	0.52	0.41
	18	0.80	0.62
	19	0.71	0.47
	20	0.63	0.59
	21	0.93	0.77
	22	0.66	0.47
	23	0.58	0.80
	24	0.58	0.73
	25	0.77	0.61
	26	0.59	0.49
	27	0.58	0.61
	28	0.63	1.00
	29	0.57	0.67
	30	0.73	0.57

table continued ➢

	Sample	Sheltered	Exposed
Sum of values		19.851	17.601
Mean		0.66	0.59
SD		0.124	0.16
Variance		0.015	0.025
Difference between means		0.075	
Number of samples		30.00	30.00
Mean variance		0.0005	0.0008
Joint variance		0.0014	
t		2.04	

Critical value of t at 5% = 2.042, therefore calculated value equals critical so reject null hypothesis.

Table 7.8 *t-test: Comparison of all the limpets on sheltered beach with those on exposed beach*

	Sample	Sheltered	Exposed
	1	0.38	0.41
	2	0.6	0.47
	3	0.39	0.51
	4	0.53	0.29
	5	0.58	0.41
	6	0.43	0.38
	7	0.47	0.57
	8	0.45	0.49
	9	0.43	0.59
	10	0.47	0.49
	11	0.55	0.53
	12	0.51	0.43
	13	0.56	0.49
	14	0.64	0.34
	15	0.49	0.53
	16	0.53	0.53
	17	0.51	0.49
	18	0.48	0.6
	19	0.41	0.55

table continued

	Sample	Sheltered	Exposed
	20	0.57	0.58
	21	0.41	0.43
	22	0.49	0.44
	23	0.47	0.43
	24	0.51	0.56
	25	0.43	0.47
	26	0.49	0.53
	27	0.53	0.48
	28	0.53	0.48
	29	0.61	0.49
	30	0.51	0.53
	31	0.71	0.44
	32	0.44	0.41
	33	0.46	0.47
	34	0.51	0.68
	35	0.68	0.44
	36	0.80	0.53
	37	0.91	0.35
	38	0.31	0.57
	39	0.37	0.23
	40	0.43	0.69
	41	0.34	0.52
	42	0.42	0.53
	43	0.58	1.02
	44	0.47	0.50
	45	0.45	0.53
	46	0.51	0.43
	47	0.40	0.56
	48	0.32	0.41
	49	0.38	0.73
	50	0.43	0.82
	51	0.49	0.40

table continued ➢

	Sample	Sheltered	Exposed
	52	0.52	0.41
	53	0.40	0.53
	54	0.35	0.47
	55	0.39	0.32
	56	0.39	0.52
	57	0.43	0.61
	58	0.33	0.41
	59	0.61	0.49
	60	0.51	0.53
	61	0.71	0.44
	62	0.44	0.41
	63	0.46	0.47
	64	0.51	0.68
	65	0.68	0.44
	66	0.80	0.53
	67	0.91	0.35
	68	0.31	0.57
	69	0.37	0.23
	70	0.43	0.69
	71	0.34	0.52
	72	0.42	0.53
	73	0.58	1.02
	74	0.47	0.50
	75	0.45	0.53
	76	0.51	0.43
	77	0.40	0.56
	78	0.32	0.41
	79	0.38	0.73
	80	0.43	0.82
	81	0.49	0.40
	82	0.52	0.41
	83	0.40	0.53

table continued ➢

	Sample	Sheltered	Exposed
	84	0.35	0.47
	85	0.39	0.32
	86	0.39	0.52
	87	0.43	0.61
	88	0.33	0.41
	89	0.61	0.49
	90	0.51	0.53
Sum of values		43.64	45.62
Mean		0.48	0.51
SD		0.122	0.132
Variance		0.015	0.018
Difference between means		0.022	
Number of samples		90.00	90.00
Mean variance		0.0002	0.0002
Joint variance		0.0004	
t		1.16	

Value of t does not exceed critical value.

Table 7.9 *t-test: Upper shore comparison of heights on sheltered and exposed beaches*

	Sample	Sheltered	Exposed
	1	26	39
	2	41	29
	3	29	33
	4	40	34
	5	32	35
	6	31	33
	7	34	29
	8	23	33
	9	46	39
	10	31	28
	11	38	31

table continued

	Sample	Sheltered	Exposed
	12	23	37
	13	30	40
	14	30	36
	15	23	34
	16	36	44
	17	24	34
	18	36	28
	19	32	33
	20	35	39
	21	37	43
	22	29	23
	23	35	36
	24	23	33
	25	30	33
	26	32	34
	27	22	39
	28	35	49
	29	32	39
	30	33	32
Sum of values		948.00	1049.00
Mean		31.60	34.97
SD		5.922	5.314
Variance		35.076	28.240
Difference between means		3.367	
Number of samples		30.00	30.00
Mean variance		1.1692	0.9413
Joint variance		2.1105	
t		2.32	

Value of t exceeds critical value, significant at the 5% level.

Table 7.10 *t-test: Exposed beach comparison of height between middle and upper shores*

	Sample	Sheltered	Exposed
	1	24	39
	2	23	29
	3	25	33
	4	38	34
	5	26	35
	6	25	33
	7	19	29
	8	25	33
	9	14	39
	10	40	28
	11	28	31
	12	24	37
	13	50	40
	14	28	36
	15	31	34
	16	23	44
	17	30	34
	18	24	28
	19	32	33
	20	47	39
	21	23	43
	22	22	23
	23	31	36
	24	23	33
	25	18	33
	26	33	34
	27	34	39
	28	23	49
	29	23	39
	30	29	32

table continued ➢

	Sample	Sheltered	Exposed
Sum of values		835.00	1049.00
Mean		27.83	34.97
SD		7.944	5.314
Variance		63.109	28.240
Difference between means		7.133	
Number of samples		30.00	30.00
Mean variance		2.1036	0.9413
Joint variance		3.0450	
t		4.09	

Value of *t* exceeds critical value, significant at the 0.1% level so reject null hypothesis.

Table 7.11 *t-test: Sheltered beach comparison of height between middle and upper shores*

	Sample	Sheltered	Exposed
	1	30	26
	2	24	41
	3	27	29
	4	23	40
	5	40	32
	6	48	31
	7	40	34
	8	18	23
	9	21	46
	10	23	31
	11	17	38
	12	19	23
	13	35	30
	14	23	30
	15	20	23
	16	30	36
	17	23	24
	18	18	36
	19	21	32

table continued ➢

	Sample	Sheltered	Exposed
	20	24	35
	21	26	37
	22	24	29
	23	18	35
	24	19	23
	25	23	30
	26	24	32
	27	24	22
	28	19	35
	29	30	32
	30	23	33
Sum of values		754.00	948.00
Mean		25.13	31.60
SD		7.343	5.922
Variance		53.913	35.076
Difference between means		6.467	
Number of samples		30.00	30.00
Mean variance		1.7971	1.1692
Joint variance		2.9663	
t		3.75	

Value of t exceeds critical value, significant at the 0.1% level so reject null hypothesis.

Table 7.12 *t-test: Comparison between the ratios of the upper shores of sheltered and exposed beaches*

	Sample	Sheltered	Exposed
	1	0.58	0.89
	2	0.73	0.52
	3	0.53	0.83
	4	0.71	0.76
	5	0.73	0.63
	6	0.63	0.70

table continued ➢

	Sample	Sheltered	Exposed
	7	0.71	0.53
	8	0.47	0.67
	9	0.82	0.68
	10	0.53	0.64
	11	0.95	0.52
	12	0.51	0.82
	13	0.67	0.71
	14	0.67	0.64
	15	0.51	0.61
	16	0.84	0.79
	17	0.52	0.61
	18	0.80	0.62
	19	0.71	0.65
	20	0.63	0.72
	21	0.93	0.77
	22	0.66	0.47
	23	0.58	0.80
	24	0.58	0.73
	25	0.77	0.61
	26	0.59	0.67
	27	0.58	0.66
	28	0.63	1.00
	29	0.57	0.87
	30	0.73	0.57
Sum of values		19.85	20.67
Mean		0.66	0.69
SD		0.124	0.121
Variance		0.015	0.015
Difference between means		0.027	
Number of samples		30.00	30.00

table continued ➢

	Sample	Sheltered	Exposed
Mean variance		0.0005	0.0005
Joint variance		0.0010	
t		0.86	

Value less than critical value so accept null hypothesis.

The *t*-test showed that there was no significant difference between the 90 limpets living on the exposed beach compared with those on the sheltered beach. However, when I analysed the different parts of the shore, I found some differences. Of my data, the height data was the most interesting. The height data produced the significant differences. I carried out a *t*-test on the height values of all the limpets living on the upper shore of the exposed and the sheltered beaches and there was a significant difference at the 5% confidence level. Then I did the same test on the height values of limpets on the upper and middle shores of the exposed beach and on the height values for the upper and middle shore of the sheltered beach. These *t*-tests also produced a significant difference, this time at the 0.1% level. Therefore, I can reject my null hypothesis, which states there will be no difference in the mean height, diameter and ratio of height to diameter of the shell of a common limpet living in different parts of the shore or on an exposed and on a sheltered beach. In fact, the limpets living on the upper shores of both beaches were taller than those limpets living lower down the shore. Also, the limpets living on the upper shore of the exposed beach had taller shells than those of the upper shore of the sheltered beach.

It is likely that the differences in shell shape, in particular the height, is due to the limpet having to be tolerant of desiccation. The higher the limpets are up the shore, the more desiccation they have to withstand. The limpets hold on tightly to the rocks to resist desiccation. The smaller the gap between their shell and the rock, the less water can escape. The more desiccation they have to resist, the tighter they hold on. Therefore, the tighter they hold on, the more they contract their foot and therefore the less space it takes up. The mantle which surrounds the foot secretes the shell. Therefore, it secretes a tall shell. The limpets on the exposed beach have to withstand the buffeting of the waves as well and this too makes them hold on tighter to the rocks with their foot. This accounts for the taller shells on the upper shore of the exposed beach, compared with the shell shape on the sheltered beach. Lower down the shore, the limpets have to resist less desiccation. This means that their feet do not need to contract as much. The mantle secretes a much wider and flatter shell.

The data on diameter and ratios did not produce any significant differences. The differences in the values were all very small and, by the time I had calculated the ratio, these differences were even smaller and were not picked up by the *t*-test. As well as the *t*-test, I worked out the standard deviation on the data. The values were quite large, indicating that I had a wide range of values.

I did look at my data for anomalous results. I noticed that the values for the ratios had a normal distribution. But some of the data sets for the lower and middle shores for height and diameter had quite a skewed distribution. My pilot test had indicated that the data would be normal and on this basis I carried out a *t*-test. However, I may have been wrong and perhaps a Mann–Whitney *U* test would have been better. Lack of time prevented me from carrying out additional tests.

Conclusion

The null hypothesis can be rejected. Limpets living on the higher shores of the beach have a taller shell in order to withstand more desiccation. Also, limpets living on the upper shore of the exposed beach have taller shells than the limpets of the same shore on the sheltered beach. This may be due to larger waves hitting the beach, and so the limpets have to hold tighter to the rocks.

Limitations

The use of the two short rulers made it easier to measure the height of the shells. However, cheap plastic rulers are not particularly accurate and so the measurements may have been out by a millimetre. I am far more confident with the measurements obtained using the vernier caliper, which gave me more precision. The shells were smooth and slippery, which made it awkward to make the measurements. Some of the limpets were located on the sides of rocks or in awkward positions for measurement. However, I still believe that the measurements that I produced were as accurate as I could make them in the circumstances.

Sampling was systemmatic, but in a few cases I had to move well away from my tape, due to the presence of large rock pools. Some of the rocks were large and those limpets sampled near the top of these rocks would have been a metre or so higher than other limpets sampled in the same zone. This may have meant that these limpets are experiencing different environmental conditions. It is also possible that the limpets that were located on the sides of rocks, sheltered from the sun and waves, may have had to withstand less desiccation than those located in more exposed positions.

Dividing the beach up into shores was done on the basis of height above sea level. However, in practice it was more difficult to determine the heights due to the presence of large rocks along the transect that I had chosen.

Some of the samples did produce skewed results, which was unexpected. Larger sample sizes may have overcome this problem.

Further study

I would like to compare the height and diameter of limpets found on the exposed and sheltered sides of a large rock, or above and within a rock pool. These differences may be more important than the overall exposure of the beach. The blue-rayed limpet is found along this beach too. I would like to see if the same differences are found in this limpet. I could also extend the study to include other species of mollusc on the shore, such as the periwinkle.

References

Adds, Larkcom, Miller and Sutton, *Tools, Techniques and Assessment in Biology*, Nelson 1999

Barrett and Yonge, *Collins Guide to the Sea Shore*, Collins 1980

Cadogan and Sutton, *Maths for Advanced Biology*, Nelson 1994

Guide to Statistics, sheets produced by Field Studies Council

Hayward, Nelson Smith, and Shields, *Collins Pocket Guide to the Sea Shore of Britain and Northern Europe*, HarperCollins 1996

Teacher's comments

Planning

There was a testable hypothesis, although this was really two linked together – one looking at differences in shell shape in different points on the beach and the second comparing two beaches that differed in exposure. However, it was possible to test both of these predictions.

The apparatus was rather basic, but appropriate for the study, and the use of the vernier caliper was justified. The number of samples and the manner of selection was described. Careful consideration was made of the choice of statistical test. A pilot study was carried out to ensure that the data had a normal distribution. There was some consideration of how the variables would be controlled. In this type of investigation, there are several variables which cannot be controlled, but mention was made of this. The method of measurement was altered in light of difficulties encountered during the pilot. The proposed method would have produced accurate and reliable data. The student made a risk assessment and thought carefully about how the investigation could affect the living organisms found on the shore.

Introduction

The biological nature of the problem was explained and linked to the hypothesis. More information could have been provided on the length of time the limpets would be exposed to the elements. Limpets living on the upper shore would have been out of water for many days at a time.

Implementation

The use of the vernier caliper takes some skill, and additional measurements were made while profiling the shores. SI units were used throughout. The investigation appeared to be organised and due care was taken of safety on the shore. The number of samples, 30 in each zone, was sufficient for the statistical test.

Introduction

The student has clearly identified the nature of the problem and supported the statement with some biological information.

Method

The method was clear and written in continuous prose. There was a logical sequence of events and considerable detail, which would enable somebody else to carry out the investigation.

Analysis

The results were fully analysed, with tables of frequency, means, standard deviations and *t*-tests. The appendix (not published here) contained tables of raw data and standard deviations. The graphs were appropriate to the study and were displayed using the correct axes and scales.

The overall trends were commented upon. The variability of the data was determined by calculating the standard deviation. This was briefly mentioned in the discussion. There was discussion of anomalous results in relation to the skewed data. The *t*-tests were laid out clearly and the use of the 5% confidence limit was appropriate. The explanations of the results were sound and were related to biological knowledge. Correct terms, such as desiccation, were used.

➢

Discussion and evaluation

The student has provided an explanation of the results and has linked this to biological knowledge. The variability of the data has been considered via the calculation of the standard deviation. The problem with the skewed data has also been raised and anomalous results have been mentioned. There is a full consideration of the limitations of the experimental technique and there are proposals which would improve the reliability of the investigation.

Style

The style was in the form of a scientific paper with an abstract that described the main features of the investigation. It was well organised with a list of references, but no cross-link to where they were used in the text. The student has made good use of specialist vocabulary.

Overall, this is good investigation and it should score quite highly. However, there are a few weaknesses so this would probably not get maximum marks for all the sections.

CHAPTER EIGHT

AQA (A and B)

AQA runs two specifications, A and B. The assessment requirements of these two specifications are different, so you need to check which one you are following.

8.1 AQA A

The AS coursework is Module 4. Students are assessed on ten skills, named A – J, which cover planning, implementation, analysis and evaluation. These skills are either assessed separately as part of the normal practical work carried out during the course, or together as part of a longer investigation. The assessment can be repeated and the best mark submitted for assessment. Within each skill, there are four performance levels, 0, 1, 2 and 3.

The A2 coursework is Module 8B. This coursework assesses seven skills, which are assessed together in a single piece of individual investigation.

The skills that are assessed are shown in Table 8.1.

Table 8.1

	Skill	Description
AS level		
Planning	A	Method of changing the independent variable.
	B	Method of measuring the dependent variable.
Implementing	C	Implementing practical work. A high performance level would show the ability to work methodically and safely, show competence in manipulative skills and managing time efficiently.
	D	Collection and presentation of raw data, with an appropriate level of accuracy. The data should be recorded in suitable form such as a table.
Drawing	E	Drawing a biological specimen, either a whole specimen, a low-power plan or a high-power drawing of a few cells as observed using a microscope. A high performance would be one with few inaccuracies and no artistic embellishments. Only the drawing is assessed, not the labelling.
Analysing	F	Use of graphical techniques, to select relevant data to produce a summary of the results.

table continued ➢

	Skill	Description
AS level		
	G	Interpretation of results with an appropriate description of the trends and patterns shown by the data, accompanied by detailed conclusions relating to biological knowledge.
	H	Evaluation of practical work. This should be more than just a list of errors which should be avoided.
Communicating		The selection and retrieval of biological information. The ideal presentation is a piece of written work of between 500 and 1000 words, that shows understanding of ethical and social, economic and environmental, technological applications of biology.
	I	The selection of appropriate information from a range of sources.
	J	Communication of biological information and the ability to present an argument logically and coherently and to express ideas with appropriate scientific terminology.
A2 level		
Planning	A	Defining the problem. The final report needs to have a title that is accompanied by a precise definition of the problem to be investigated. The introduction needs to draw biological information from across the specification (i.e. it needs to be synoptic). The plan needs to be written in the future tense.
	B	Changing an independent variable, with consideration as to how the other variables would be controlled.
	C	Method of measuring the dependent variable to produce a full range of reliable quantitative data.
Implementation	D	Implementing practical work. A high performance level would show the ability to work methodically and safely, show competence in manipulative skills and manage time efficiently.
Analysis of evidence and drawing conclusions	E	This involves the use of statistical techniques. This needs to be an acceptable technique and the calculation complete and accurate.
	F	The interpretation of results needs to include an appropriate description of the trends and patterns shown by the data and should be accompanied by detailed conclusions relating to biological knowledge drawn from across the specification.

table continued ➢

	Skill	Description
AS level		
Evaluating and communicating	G	Evaluating evidence and procedures. This should include the main sources of error with information on how anomalous results were dealt with.

8.2 AQA B

The AS coursework consists of four elements: planning, implementation, analysis and evaluation. These can be carried out separately. Each element is worth between 2 and 8 marks. At AS level, the coursework module is Unit 3(b) and at A2 the coursework is Unit 5(b) (Table 8.2).

AS level

Table 8.2

	The candidate can :
Planning	
2 marks	identify the problem, suggesting relevant procedures to be used and selecting appropriate equipment and materials to carry out the task.
4 marks	design a plan in which enough relevant factors are taken into account and controlled in order to obtain valid data.
6 marks	describe the observations, measurements and precautions needed to obtain valid and where appropriate quantitative data.
8 marks	provide a reasoned explanation for the procedures selected, describe the anticipated method for collecting the data and suggest how they might be analysed.
Implementing	
2 marks	safely use appropriate techniques and equipment to obtain and record some relevant observations and measurements, with units given when applicable.
4 marks	use techniques and equipment in a methodical and organised way to obtain and record an adequate range of valid observations and measurements.
6 marks	use appropriate techniques and equipment to make detailed observations and suitably accurate quantitative measurements and record these in clear and understandable form.
8 marks	take precautions to ensure the reliability of the data obtained in relation to the problem being investigated.

table continued ➢

	The candidate can :
Analysis	
2 marks	carry out some processing of results, for example, in the form of a simple graph, chart or diagram and identify the main pattern or trend.
4 marks	process the data obtained with the appropriate graphs and calculations, describe the relevant trends and patterns and provide a conclusion or solution to the problem which is consistent with the data.
6 marks	using appropriate scientific terminology, to provide a full report of the data obtained, analysing in detail, patterns within the data obtained and draw appropriate conclusions.
8 marks	make a full analysis of the outcomes of the investigation, supporting these with appropriate evidence from the data and explaining conclusions in relation to scientific knowledge.
Evaluation	
2 marks	recognise the limitations of the apparatus and the techniques used.
4 marks	assess the reliability and precision of the data obtained.
6 marks	assess the effect of the limitations of the apparatus, techniques and reliability on the conclusions made.

A2 level

A2 coursework is one piece of coursework covering all four skill areas carried out at the same time (Table 8.3).

Table 8.3

	The candidate can :
Planning	
2	clarify the problem suggesting the relevant procedures to be used and selecting appropriate equipment to carry out the task.
4	design a plan in which enough relevant factors are taken into account and controlled in order to obtain valid data.
6	describe the observations, measurements and precautions needed to obtain valid quantitative data.
8	provide a reasoned explanation for the procedures selected, describe the anticipated method of collecting the data and suggest how they might be analysed, including an appropriate simple statistical test to support or reject the hypothesis being tested.

table continued ➢

	The candidate can :
Implementation	
2	safely use appropriate techniques and equipment to obtain some relevant observations and measurements, with units given where applicable.
4	use techniques and equipment to obtain and record an adequate range of valid observations and measurements.
6	use appropriate techniques and equipment to make detailed observations and suitably accurate quantitative measurements and record these in a clear and understandable form.
8	take precautions to ensure the reliability of the data obtained in relation to the problem being investigated.
Analysis	
2	carry out some processing of results, e.g. in the form of a simple graph, chart or diagram, and identify the main pattern or trend.
4	process the data obtained with appropriate graphs and calculations, describe the relevant trends and patterns and provide a conclusion or solution to the problem, which is consistent with the data.
6	using appropriate scientific terminology, provide a full report of the data obtained, analyse in detail patterns within the data obtained and draw appropriate conclusions.
8	make full analysis of the outcomes of the investigation, supporting these with appropriate evidence from the data and explain conclusions in relation to scientific knowledge.
Evaluation	
2	recognise the limitations of the apparatus and techniques used.
4	assess the reliability and precision of the data obtained.
6	assess the effect of the limitations of the apparatus, techniques and reliability on the conclusions made.

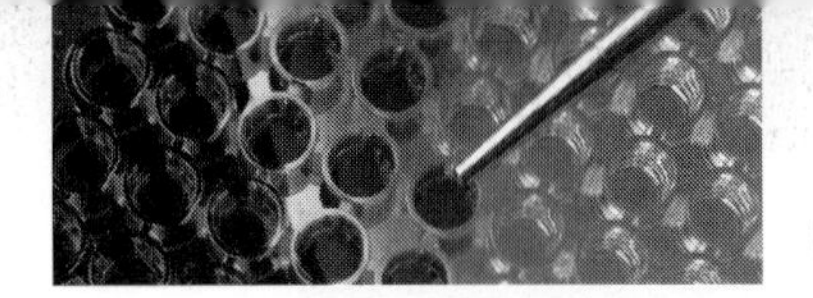

CHAPTER NINE

Edexcel

The assessment of coursework is divided into two parts, T1 and T2. T1 is part of Unit 3 and is taken by students at AS level. This coursework is assessed by teachers. T2 is part of Unit 6 and is taken by A2 students. It is assessed partly by teachers and partly by external examiners. Both papers are worth 32 marks. At AS level, T1 contributes 15% to the final subject mark. At A2, T1 and T2 together contribute 15% to the final mark. There is a written alternative paper to T2 which can be taken by A2 candidates.

9.1 AS level coursework

The total marks of 32 are allocated as shown in Table 9.1.

Table 9.1

	Marks
Planning	8
Implementation	8
Analysing evidence and conclusions	8
Evaluation	8
Total	32

The assessed coursework consists of a single written report of an individual investigation. This investigation must be quantitative, i.e. involve making measurements that yield data. It must be linked to Units 1, 2 and 3. It is expected that students will take approximately 3 to 4 hours to carry out the investigation and they are expected to work on their own.

9.2 A2 level coursework

This second piece of coursework involves a written report of an individual study. This coursework involves students planning and carrying out a scientific investigation and writing a report of the practical work in a style that is similar to that used in scientific journals. It has to be a quantitative investigation and involve the use of an appropriate statistical test. It can cover any of the topics in the AS and A2 specifications. The report should be approximately 3500 words (see Table 9.2).

Table 9.2

	Marks
Planning	4
Implementation	4
Introduction	4
Methods	3
Analysing evidence	6
Discussion and evaluation	8
Style	3
Total	32

The marking of this piece of coursework is slightly different to that of AS. Your teacher will assess the planning and implementation of the study and then the report is sent into Edexcel for marking by an examiner.

9.3 A2 written alternative test

Some schools may allow their students to sit the written alternative to T2 coursework. This examination is still worth 32 marks and it is based on practical questions. You will be asked to plan theoretical investigations and manipulate experimental data. You will be tested on the skills that you would have used in carrying out an individual investigation.

9.4 Detailed criteria

AS assessment Unit 3 (Table 9.3)

Table 9.3

Marks	Planning
2	(a) A testable hypothesis is formulated with guidance and either the biological knowledge used to explain the nature of the problem is superficial or help is needed to select suitable knowledge. (b) The plan produced is based on previously encountered or familiar procedures or further guidance is needed to develop this into a worthwhile investigation which is safe. (c) Awareness of the acceptable treatment of living organisms and safe practice is limited or advice is needed.

table continued ➢

Marks		Analysing evidence and drawing conclusions
4	(a)	Summary tables of the observations and calculations are presented to illustrate trends in the data. Graphs are drawn to display these trends, the graphs are suitably labelled with the correct choice of axis for each variable.
	(b)	Trends and patterns in the data are recognised.
	(c)	Conclusions are drawn and explanations of experimental results are related to basic biological knowledge and understanding.
6	(a)	Summary tables of the observations and calculations are presented to illustrate trends in the data. Selective choice of graph, displays important trends and patterns in the correct format using SI units appropriately.
	(b)	Trends and patterns in the data are clearly recognised and commented on. Some anomalies or inconsistencies are indicated.
	(c)	Explanations of experimental results are sound and clearly related to biological knowledge and understanding.
8	(a)	A high degree of competence is shown in the presentation and tabulation of the collected data. Appropriate graphs are carefully chosen to display the important trends, patterns and comparisons. There is use of the correct format including error bars and there is no undue repetition. SI units are used accurately at all times.
	(b)	Trends and patterns in the data are clearly recognised and commented on. All apparent anomalies and inconsistencies are described.
	(c)	Coherent, logical and comprehensive explanations of experimental results are given using carefully selected, appropriate biological knowledge and terminology.

Marks		Evaluating evidence and procedures
2	(a)	Analysis of the variability of results and the reliability of conclusions is very limited or considerable help is needed to make relevant comments.
	(b)	Difficulties with apparatus or measurements are described. Simple suggestions are made for further investigations or repeated measurements.
4	(a)	Variability of results and apparent anomalies are discussed. Limited comments are made on the reliability of the conclusions drawn or the variation between expected and actual results.
	(b)	Comments are made on difficulties encountered when collecting data or handling apparatus. There is some attempt to explain how these may have affected the results. Reasonable suggestions are made for improved techniques.

table continued ➢

Marks	Evaluating evidence and procedures
6	(a) Variability of results and apparent anomalies are used to assess the reliability and precision of the experimental data and the conclusions drawn from them. (b) Limitations of the experimental techniques employed are discussed. Proposed suggestions for further investigations would provide some additional evidence for the conclusion or extend the enquiry.
8	(a) Variability of results and apparent anomalies are used to assess the reliability and precision of the experimental data and the conclusions drawn from them. The critical analysis shows an awareness of the tentative nature of the results of single investigations. (b) Limitations of the experimental techniques employed and their influence on the results are discussed in detail. Proposed suggestions for further investigations would provide considerable additional evidence for the conclusions or extension to the enquiry.

A2 assessment Unit 6 (Table 9.4)

Table 9.4

Marks	Planning
2	(a) A testable hypothesis is formulated independently but may require some modification. (b) Apparatus and procedures to be used are clearly described and the number and type of measurements or observations are adequate to generate useful and reliable results. Some important variables are identified and controlled to allow for the collection of meaningful results. (c) The suggested procedure can be carried out safely without the need for further guidance and involves acceptable treatment of living organisms and the environment.
4	(a) A testable hypothesis is formulated independently and is stated in a concise form. The planned approach shows some originality even where familiar problems are investigated. (b) The choice of apparatus and materials is fully justified. It is clear how the number and type of measurements or observations were chosen to generate useful and reliable results and provide suitable data for analysis by a named statistical test. The planned procedures describe clearly how all important variables are to be controlled to produce reliable results. (c) Thorough risk assessments of hazardous procedures or substances are undertaken and full consideration is given to the ethical implications of the choice of living organisms and the environment.
	table continued ➢

Marks		Implementing
2	(a)	Apparatus and materials are handled competently. A range of manipulative techniques are used safely with some skill.
	(b)	The investigation is sufficiently well organised to allow it to be completed without assistance. The work is carried out safely with sufficient regard for the well being of living organisms and the environment.
	(c)	Measurements are made with reasonable precision. Original observations are recorded in a structured manner using suitable tables with headings and units.
4	(a)	Apparatus and materials are handled competently. A wide range of manipulative techniques are used safely and with a high degree of skill.
	(b)	The investigation is well organised and carried out in a methodical fashion with meticulous attention to safety at all times. Due consideration is given to the well being of living organisms and the environment.
	(c)	Measurements are made to a high degree of precision and attention to detail. Recordings are repeated to ensure that the number and type of observations are accurately linked to the hypothesis being tested. All original observations are methodically and accurately recorded in a suitable table with headings and SI units.
Marks		**Introduction**
2	(a)	Some biological knowledge and understanding is used to explain the nature of the problem but this is not always relevant or clearly linked to the stated hypothesis.
4	(a)	The nature of the problem to be investigated is clearly defined using carefully selected and relevant biological knowledge, principles and concepts. There is a clear and logical progression from the background information used to a statement of a concise hypothesis.
Marks		**Method**
1	(a)	The method account is disjointed or written as a list.
	(b)	It is difficult to repeat the experiment from the method described.
	(c)	Little attention is given to describing the precautions needed to ensure reliable or accurate results.
2	(a)	The method is written in continuous prose and is easy to follow.
	(b)	Some details of the methods employed are missing or the account lacks a logical sequence but it is still possible to repeat the experiment.
	(c)	There is some description of the precautions taken to ensure the accuracy and reliability of the data collected.

table continued ➢

Marks		Method
3	(a)	The method account is clear, concise and written in continuous prose.
	(b)	The account follows a logical sequence and includes sufficient detail to allow the reader to replicate the procedure.
	(c)	The account describes and explains all precautions taken to control variables and any amendments to the plan which were used to improve accuracy or reliability.

Marks		Analysing evidence
2	(a)	Summary tables of the observations and calculations are presented to illustrate trends in the data. Graphs are drawn to display these trends, the graphs are suitably labelled with the correct choice of axis for each variable.
	(b)	Trends and patterns in the data are recognised.
	(c)	Statistical analysis is absent or is only completed with detailed guidance. Application of calculated statistical values is present, though limited or confused.
4	(a)	Summary tables of the observations and calculations are presented to illustrate trends in the data. Selective choice of graph displays important trends and patterns in the correct format using SI units appropriately.
	(b)	Trends and patterns in the data are clearly recognised and commented on. Some anomalies or inconsistencies are indicated.
	(c)	The chosen statistical test may be inappropriate or provide limited analysis of the stated hypothesis. Calculations are clearly set out but the interpretation of calculated values lacks detailed explanation.
6	(a)	A high degree of competence is shown in the tabulation and presentation of the collected data. Appropriate graphs are carefully chosen to display the important trends, patterns and comparisons. There is use of the correct format including error bars and there is no undue repetition. SI units are used accurately at all times.
	(b)	Trends and patterns in the data are clearly recognised and commented on. All apparent anomalies and inconsistencies are described.
	(c)	The chosen statistical test is appropriate to the data to be analysed and the hypothesis to be tested. Calculations of statistical tests are clearly set out and interpreted using a null hypothesis and 5% confidence levels where appropriate.

table continued ➢

Marks	Discussion and evaluation
2	(a) Some superficial conclusions which make limited reference to basic biological knowledge and understanding. Help may be needed to relate the conclusion to biological knowledge and understanding. (b) There is a limited analysis of the variability of results and the reliability of conclusions and with considerable help comments are made. (c) Difficulties with apparatus or measurements are described. Simple suggestions are made for further investigations or repeated measurements.
4	(a) Conclusions are drawn and explanations of experimental results are related to basic biological knowledge and understanding. (b) Variability of results and apparent anomalies are discussed. Limited comments are made on the reliability of the conclusions drawn or the variation between expected and actual results. (c) Comments are made on difficulties encountered when collecting data or handling apparatus. There is some attempt to explain how these may have affected the results. Reasonable suggestions are made for improved techniques.
6	(a) Explanations of experimental results are sound and clearly related to biological knowledge and understanding. (b) Variability of results and apparent anomalies are used to assess the reliability and precision of the experimental data and the conclusions drawn from them. (c) Limitations of the experimental techniques employed are discussed. Proposed suggestions for further investigations would provide some additional evidence for the conclusion or extend the enquiry.
8	(a) Coherent, logical and comprehensive explanations of experimental results are given using carefully selected, appropriate biological knowledge and terminology. (b) Both sides of an argument are presented clearly and concisely when evidence from the investigation is weighed up. Variability of results and apparent anomalies are used to assess the reliability and precision of the experimental data and the conclusions drawn from them. The critical analysis shows a clear awareness of the tentative nature of the results of single investigations. (c) Limitations of the experimental techniques employed and their influence on the results are discussed in detail. Proposed suggestions for further investigation would provide considerable additional evidence for the conclusions or extension to the enquiry.

table continued ➢

Marks	Style
1	(a) The account of the investigation is present, but lacks a detailed abstract. (b) References may be quoted but there is little evidence of their application in the account of the investigation. (c) Vocabulary is limited and there are numerous errors of spelling, punctuation or grammar.
2	(a) The account of the investigation is preceded by an abstract, which describes the main features of the investigation. (b) Some references are listed but it is not clear how these have been used. There is some evidence of over elaboration or lack of organisation. (c) Specialist vocabulary is used, but there are some errors of spelling, punctuation or grammar.
3	(a) The account of the investigation is preceded by a concise abstract describing the main aims, methods, results and summarising the main conclusions. (b) References to sources used are clearly indicated in the body of the text in an appropriate manner. The whole account is concise and well organised. (c) Good use is made of specialist vocabulary and there is accurate use of spelling, punctuation and grammar.

Written alternative test

This written examination is an alternative to the T2 coursework. It tests the same skills, such as planning, implementation, analysis and evaluation. The examination is 1 hour 20 minutes long and consists of two questions worth a total of 32 marks.

Question 1 is designed to test your ability to organise and interpret data. You will usually be given some raw data and asked to tabulate it, then plot a graph, suggest a suitable statistical test and draw conclusions. The tabulation may require you to prepare a tally chart or produce a table of derived data by carrying out a calculation such as percentage change or a mean. You will not be expected to be able to carry out a statistical test from memory, but to be able to suggest which test would be appropriate and how you would interpret the test in terms of confidence limits.

Specimen question

Many fruits and seeds contain germination inhibitors, which delay germination until the inhibitors have been washed away by rain or become inactivated. In tomatoes, the inhibitors delay germination until the fleshy tomato fruit has rotted away releasing the seeds. It has been suggested that, in tomatoes, the inhibitor of seed germination is present in the fleshy tissues of the fruit rather than in the seed coats. Plan an investigation, which you could personally carry out, to test this hypothesis. Your answer should give details under the following headings.

(a) Plan of the investigation to be carried out. (9 marks)
(b) Recording the raw data measurements, presentation of results and methods of data analysis. (7 marks)
(c) Limitations of your method and an indication of further work which could be undertaken. (5 marks)

Total 21 marks.

Mark scheme

(; at end of row indicates one mark point.)

(a) Seeds from same species/variety;
Reference to optimum conditions for growth of seeds;
Oxygen/air/water;
Suitable temperature (e.g. 15–20°C);
Physical conditions for germination kept constant;
Correct substrate for growth (soil/compost/cotton wool, etc.);
Remove seeds from ripe tomato;
Test for germination/count number germinated;
Test for germination if flesh present;
Wash tomato flesh in running water;
Use this in medium to test for germination or plant seeds with washed flesh;
Plant seeds with seed coat extract or remove seed coat;
Plant seeds without seed coat extract or with seed coat;
Check for germination in thoroughly washed seeds;
Germination defined, for example appearance of radicle;
Large number of seeds in each test;
Suitable replicates;
Calculation of rate of germination percentage or equivalent;

Total 9 marks.

(a) Format of table;
Columns of washed and unwashed flesh;
Columns for dried seeds;
Column for % germination/times for germination;
Extra boxes for repeats;
Bar chart/line graph;
Correct axes;
Calculation of means;
Calculation of standard deviations;
Use of *t*-test;
Interpretation of *t*-test;

Total 7 marks.

(b) Tomatoes vary in ripeness;
Initial treatment or conditions could vary;
Assumes inhibitor is water soluble or would be removed by washing;
Effects could be caused indirectly by a form of inactivation;
Test effects on different strains;
Test for seasonal variation;
Test effects of extracts alone;

Total 5 marks.

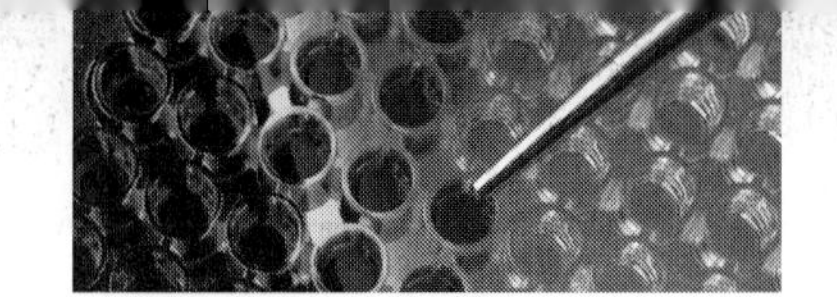

CHAPTER TEN

OCR (Oxford, Cambridge and RSA Examinations)

OCR offers course work modules at AS and A2. In addition, there is the option for candidates to take an externally assessed practical examination.

The combinations that are possible are:

- AS and A2 coursework assessed by teacher in own centre
- AS and A2 assessed by externally assessed practical examination
- AS coursework assessed by centre and A2 practical examination assessed externally
- AS practical examination assessed externally and A2 coursework assessed by centre.

At AS level the practical assessment is part of Unit 2803. Unit 2803/02 is centre-assessed coursework and Unit 2803/03 is the practical examination. At A2 level, the practical assessment is Unit 2806. Unit 2806/02 is the coursework and Unit 2806/02 is the practical examination.

There are four skill areas (Table 10.1):

Table 10.1

Skill	Name	Marks
Skill P	Planning	8
Skill I	Implementing	7
Skill A	Analysing evidence and drawing conclusions	8
Skill E	Evaluating evidence and procedures	7

At AS and A2 level there are a total of 30 marks available for course work. The final mark is doubled to give a total mark out of 60. The assessment for each skill can be repeated both at AS and A2 and the best mark submitted.

10.1 Detailed assessment criteria

The statements given in bold are additional A2 requirements.

Skill P - Planning: total 8 marks (Table 10.2)

Table 10.2

Mark	Descriptor	The candidate:
1	P.1a	develops a question or problem in simple terms and plans a fair test or an appropriate practical procedure, making a prediction where relevant.
	P.1b	chooses appropriate equipment.
3	P.3a	develops a question or problem using scientific knowledge and understanding **drawn from more than one area of the specification**, identifies the key factors to vary, control or take account of.
	P.3b	decides on a suitable number and range of observations and/or measurements to be made.
5	P.5a	uses detailed scientific knowledge and understanding **drawn from more than one area of the specification** and information from preliminary work or a secondary source to plan an appropriate strategy, taking into account the need for safe working and justifying any prediction made.
	P.5b	describes a strategy, including the choice of equipment which takes into account the need to produce precise and reliable evidence, produces a clear account and uses specialist vocabulary appropriately.
7	P.7a	retrieves and evaluates information from a variety of sources and uses it to develop a strategy which is well structured, logical and linked coherently to underlying scientific knowledge and understanding **drawn from different parts of the AS and A2 specification**, uses spelling, punctuation and grammar correctly.
	P.7b	justifies the strategy developed, including the choice of equipment, in terms of the need for precision and reliability.
8		

Skill I - Implementing: total 7 marks (Table 10.3)

Table 10.3

Mark	Descriptor	The candidate:
1	I.1a	demonstrates competence in simple techniques and an awareness of the need for safe working.
	I.1b	makes and records observations and/or measurements which are adequate for the activity.

table continued ➢

Mark	Descriptor	The candidate:
3	I.3a	demonstrates competence in practised techniques and is able to manipulate materials and equipment with precision.
	I.3b	makes systematic and accurate observations and/or measurements which are recorded clearly and accurately.
5	I.5a	demonstrates competence and confidence in the use of practical techniques, adopts safe working practices throughout.
	I.5b	makes observations and/or measurements with precision and skill, records observations and or measurements in appropriate format.
7	I.7a	demonstrates skilful and proficient use of all techniques and equipment.
	I.7b	makes and records all observations and/or measurements in appropriate detail and to the degree of precision permitted by the techniques or apparatus.

Skill A - Analysing evidence and drawing conclusions: total 8 marks (Table 10.4)

Table 10.4

Mark	Descriptor	The candidate:
1	A.1a	carries out some simple processing of the evidence collected from experimental work.
	A.1b	identifies trends or patterns in the evidence and draws simple conclusions.
3	A.3a	processes and presents evidence gathered from experimental work including, where appropriate, the use of appropriate graphical and or numerical techniques.
	A.3b	links conclusions drawn from processed evidence with the associated scientific knowledge and understanding **drawn from more than one area of the specification.**
5	A.5a	carries out detailed processing of evidence and analysis including, where appropriate, the use of advanced numerical techniques such as statistics, plotting of intercepts or the calculation of gradients.
	A.5b	draws conclusions which are consistent with the processed evidence and links these with detailed scientific knowledge and understanding **drawn from more than one module of the specification**, produces a clear account which uses specialist vocabulary appropriately.
		table continued ➤

mammal and the other had a second section, this time cut along a different plane. They were required to mount the sample of sheep's kidney on a slide and make a low power drawing. They looked at the other two slides using a hand lens and microscope to identify certain regions. Finally, they had to make a high power drawing of a typical tubule seen in the kidney.

A2 Specimen practical paper

Question 1 planning task

Figure 1 shows a diagram of a respirometer that can be used to measure the respiratory rate of small organisms. The apparatus works in the following way.

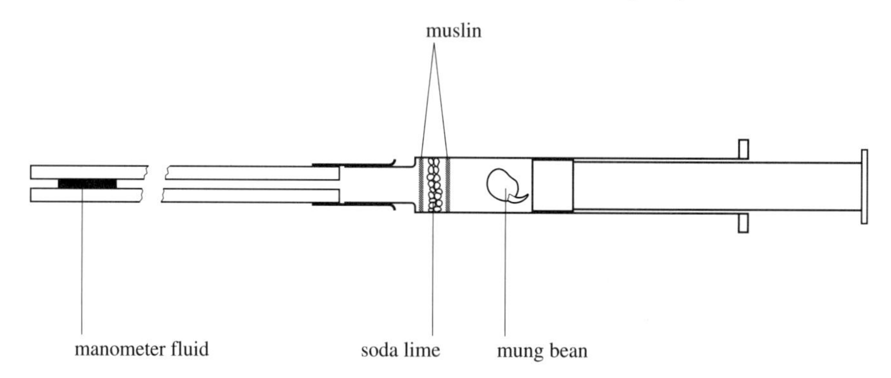

Figure 10.1

Oxygen absorbed by the respiring material, e.g. a germinating mung bean, causes a volume reduction to occur in the apparatus because the carbon dioxide given out is absorbed by the soda lime. The reduction in volume causes the manometer fluid to move towards the syringe. Movement of the manometer fluid, e.g. in millimetres per minute, provides a measure of the oxygen uptake by the respiring material. The manometer fluid can be returned to its original material by moving in the syringe plunger.

Design and describe a procedure that would allow you to compare the respiratory rate of mung bean seedlings that have been germinated for 2 days and for 4 days. In your account refer to:

- any variables you would need to consider, in addition to the age of the seedlings
- the data you would collect and how that would be done
- any procedures you would adopt to ensure that the data you collect and the deductions you draw from them are as reliable as possible
- how the data would be used to make the comparison in respiratory rate.

Question 2

Candidates were provided with plant material and four different sucrose solutions. They were asked to estimate the water potential of the cells in the plant material.

The second part of Question 2 involved a prepared microscope slide of a transverse section through a piece of plant tissue similar to that used in the investigation. Candidates had to make a high-power drawing showing (a) three adjoining epidermal cells and (b) three adjoining cells next to the central cavity of the section to the same scale.

Question 3

Candidates were supplied with a microscope slide with a stained transverse section through the small intestine of a mammal. They were asked to draw a low-power plan of the section and to calculate the magnification of the drawing. Finally, they were asked to draw a high-power drawing of three adjoining cells, followed up by some questions on the functions of these cells.

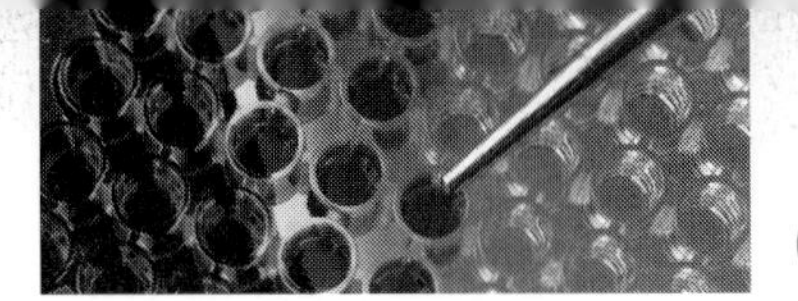

CHAPTER ELEVEN

WJEC (Welsh Joint Education Committee)

The basic assessment criteria are the same at AS and A2 levels. It is expected that students will show some progression from AS to A2, therefore, investigations at A2 should involve slightly more complex and demanding activities. For example, analysis and evaluation at AS level involves graphical presentation of data, whilst at A2 level a statistical analysis is required. Microscope work at AS level involves calibration of the microscope and observation of the specimen at low power or plan level. At A2 level this usually involves the examination and drawing of a section at high power, along with further related investigative activities.

11.1 Scheme of work

Candidates are required to work through a specific scheme of assessment provided by the WJEC during the spring term. The planning, analysis and evaluation skills at both AS level and A2 level are assessed in a series of 1 hour practical sessions set on the same day or in blocks. The assessment scheme may require the planning and implementing of one investigation or the implementing of a different investigation. The planning of the A2 investigations include a statistical t test and are conducted under open book, supervised conditions within the 1 hour time limit. If the planning and implementing involve the same investigation, the planning part of the assessment may be carried out in a 1 hour session and then implemented in another session. In between the sessions, the plans are retained by your teacher. Small adjustments to the plan are permissible as points may become apparent during the implementing session, but any amendments will be annotated as being later adjustments. If a different investigation is used for implementing, the planning work is retained by your teacher with no further amendment. The completed work is submitted to an external assessor for marking during the summer term.

11.2 Detailed assessment criteria

AS and A2 assessment planning

(the marks for each section are given in [ ])

1 Aim of the investigation – a quantitative statement [2]

2 Biological theory/model clearly explained [2]

3 Selection of apparatus and listed [2]

4 Drawing of the apparatus set up [2]

5 Variables identified:

(a) Independent and range to be used [1]

(b) Dependent [1]

(c) Controlled variables [2]

6 Suitable experimental control [1]

7 Risk assessment – the nature of the risk and how it will be minimised [2]

8 Logical sequence of the steps involved:

(a) The steps must be repeatable and the appropriate units of measurement used [5]

(b) Precision of measurement [1]

(c) Suitable repeats [2]

(d) Clear scientific account [2]

Total 25 marks

AS and A2 assessment observation and recording of results

1 Suitable table with suitable title and correct headings [2]

2 Recording accuracy – precision commensurate with instruments used (e.g. ± 1 mm) and sufficient repeats (3 – 5) plus means recorded [2]

Total 4 marks

AS and A2 assessment implementing

This skill is assessed by your teacher over a number of practical sessions. The skills must be at the appropriate level for either AS or A2. Your teacher will supply the board with a list of the practicals that have been assessed.

The marks are allocated as follows:

1 Carrying out the practical in a careful and organised way [2]

2 Manipulative skills:

(a) Setting up the apparatus correctly [2]

(b) Precise manipulation of key instruments [2]

Total 6 marks

AS level assessment analysis and evaluation

1 The processing of data in a suitable format, i.e. graphs as appropriate.

(a) Correct axes and units with the independent variable on the x-axis [2]

(b) Accurate plotting of plots [2]

(c) The drawing of a line or curve of best fit (if sufficient data are available). Otherwise the joining of plots with no extrapolation [2]

(d) Suitable scales on the graph [2]

2 Error bars drawn on the graph. Comment on the reliability of the results [2]

3 Describe the limitations of the apparatus, materials and the method and all possible sources of error in the investigation [2]

4 Draw suitable conclusions using the data [2]

5 Explanation of the conclusions using biological knowledge. This is to be relevant, concise and accurate and to comment on the validity of the investigation [4]

6 Improvement or extension to the investigation [2]

Total 20 marks

AS assessment observation - microscopy

1 Calibration of the microscope for the appropriate objective lens [1]

2 Correct statement of eyepiece divisions equating with the correct micrometer units [1]

3 Actual calculation of one eyepiece division in the appropriate units [1]

Quality of the drawing

1 A large drawing of the correct distribution of all tissues [2]

2 Clean, single, sharp and complete lines with no shading [2]

3 Drawing the correct proportions of all tissues [2]

4 The correct identification by unambiguous labelling [2]

5 A line drawn on the drawing to show its maximum length/width. The correct measurements in mm, annotated on the line (+/− 1 mm error allowed) [2]

6 Using the values above, show how the magnification of the drawing was calculated [2]

Total 15 marks

A2 assessment analysis and evaluation

Numerical processing of results using statistical methods. Correct use of *t*-test.

1 Null hypothesis stated [1]

2 Calculate deviations from the mean [2]

3 Correct substitution into the formula [2]

4 Correct statement of degrees of freedom [1]

5 Correct method to generate a value [2]

6 Use the value generated to explain the validity of the results using confidence levels [2]

7 Conclusions, using biological principles [6]

8 Limitations of the experiment and all sources of error [2]

9 Improvements or extensions to the investigation [2]

Total 20 marks

A2 assessment observation - microscopy

1 A large drawing of the correct distribution of all tissues [2]

2 Clean, single, sharp and complete lines with no shading [2]

3 Drawing the correct proportions of all tissues [2]

4 The correct identification [2]

5 The correct measurements [1]

6 Further explanation or observations (according to context) [6]

Total 15 marks

Implementing

This refers to the assessment of the practical performance of a candidate by their ability to work in a safe, ordered manner, showing good laboratory practice and using common biological equipment in a competent, precise and skilful manner. These skills are assessed by your teacher either during the practical scheme provided by WJEC or during other practical sessions throughout the year. There is no set time for this assessment, but no practical exercise will take more than 1 hour. Three skills are assessed: carrying out the work; setting up apparatus correctly; precise manipulation of key instruments. Only one mark is required for each skill although each may be assessed on several occasions and not all need to be assessed at the same time.

The three skills should be assessed according to the performance level criteria as in Table 11.1:

Table 11.1

Skill	Performance level marks		
	0	1	2
Carrying out the work	Little care taken or organisation, with little attention to safety.	Limited care taken and organisation, with some attention to safety.	Careful and well organised, safety conscious.
Setting up apparatus correctly	Little skill shown setting up/using apparatus.	Apparatus set up/used with limited skill.	Apparatus set up/used well and correctly.
Precise manipulation or key instruments	Instruments manipulated with little precision or skill.	Instruments manipulated with limited precision and skill.	Instruments manipulated with great precision and skill.

GLOSSARY OF KEY TERMS

Abstract
A concise account or summary of an investigation describing the aims, hypothesis, methods, results and conclusions. Usually presented at the start of a scientific report.

Accuracy
The exactness or precision of a measurement. The accuracy of a measurement depends on the experimental techniques and equipment used and the skill of the experimenter. Accuracy can be improved by removing or minimising sources of error.

Anomaly
A value or measurement that is out of line or irregular compared with the other results. Anomalous results can be due to experimental error.

Conclusion
A concise explanation of a set of experimental results linked to biological knowledge.

Control
A standard of comparison for checking the results of an experiment. For example, in an enzyme experiment there may be one treatment, the control, in which there is no enzyme. This means it is possible to attribute the result of the experiment to the enzyme and not to some other factor.

Evaluation
An assessment of the reliability and precision of the observations and measurements taken during an experiment, of the limitations of the techniques and of the conclusions which have been drawn.

Exponential
Of an increase that becomes more and more rapid.

Hypothesis
A proposal based on scientific knowledge and understanding designed to explain a particular problem or set of observations and measurements. Once a hypothesis has been formulated, it is possible to make predictions and to test the predictions by carrying out an appropriate experiment. A null hypothesis is used in statistical tests. The null hypothesis is the opposite of what you expect, for example there will be no difference.

Limitations
The restrictions of a particular experimental technique or set of apparatus. Limitations encountered during an experiment could influence the results and would have to be addressed in the evaluation.

Mean
Also called average. A value that is calculated by adding up the values and dividing the sum by the number of values.

Median
The middle value in a range of results.

Mode
The most frequent measurement.

Precision
The accuracy and reliability of measurements. This is dependent on the experimental technique and the apparatus selected by the experimenter. The precision of observations and measurements may be determined by the type of investigation. For example, when measuring the length of plant shoots, it may be sufficient to measure to the nearest millimetre. It would be pointless to measure to fractions of millimetres.

Qualitative
Based on non-numerical observations, for example colour.

Quantitative
Based on numerical data.

Reliability
A measure of the confidence that can be placed in a set of observations or measurements. The reliability of a set of observations or measurements depends on the number and accuracy of the individual observations or measurements. Reliability can be improved by replicating observations and measurements.

Replicates
Repeat observations or measurements.

Risk assessment
A consideration of the possible safety hazards that could be encountered during an experiment.

Trend
The general direction, tendency or pattern shown by a set of observations or measurements.

Validity
A measure of the confidence that can be placed in a conclusion. The validity depends on factors such as the range and reliability of observations and measurements. A conclusion may relate to whether or not the hypothesis can be accepted or rejected. Statistical tests may be used to place a value on the reliability of data. The tests generate a probability that the data conform with the hypothesis.

Variable
A factor that will affect an experiment. There are two types of variable, the dependent and independent variable. The independent variable is the factor that is controlled within the experiment. The dependent variable depends on the independent variable. For example, in a catalase experiment, the temperature at which the enzyme reaction takes place is the independent variable that is being controlled and the oxygen production is the variable which is being measured. Other independent variables will have be controlled in order that they do not affect the results, for example pH and substrate concentration.

Variability
The degree to which the observation or measurements differ from one another.

BIBLIOGRAPHY

Useful Books

Adds, J, Larkcom, E, Miller, R and Sutton, R, *Tools, Techniques and Asessment in Biology*, Nelson 1999

Association of Science, *Topics in Safety*, 1988 and 1998

Cadogan, A and Sutton, R, *Maths for Advanced Biology*, Nelson 1994

Jones, A, Reed R and Weyers, J, *Practical Skills in Biology*, Addison Wesley Longman 1998

CLEAPSS School Science Service, *Laboratory Handbook* (Consortium of Local Education Authorities for the Provision of Science Services). This handbook should be found in most school laboratories and it gives guidance on using chemicals and equipment. Website www.cleapss.org.uk

DfEE, *Safety in School Science*, HMSO 1996

Roberts, M B V, Reiss, M and King, T J, *Practical Biology for Advanced Level*, Nelson 1994

Useful journals

There are a number of specialist science and biology journals that can provide ideas for investigations and advice on the presentation of reports

School Science Review

Journal of Biological Education

Biological Sciences Review

Useful documents

British Psychology Society

Code of Conduct, Ethical Principles and Guidelines

http://www.bps.org.uk/documents/Code.pdf

This downloadable booklet outlines the procedures and precautions you should take when using either humans or animals in your investigations.

Websites

Science and Plants for Schools (SAPS)

This Website provides information about the use of fast cycling brassicas in schools. There is lots of information about growing these plants and using them in investigations.

www.saps.plantsci.cam.ac.uk

INDEX

Note: page numbers in italic refer to tables and diagrams